Letters and Diaries

OF

A. F. R. WOLLASTON

A. F. R. WOLLASTON

Letters and Diaries
OF
A. F. R. WOLLASTON

Selected and edited
by
MARY WOLLASTON

With a preface by
SIR HENRY NEWBOLT

CAMBRIDGE
AT THE UNIVERSITY PRESS
1933

CAMBRIDGE
UNIVERSITY PRESS

University Printing House, Cambridge CB2 8BS, United Kingdom

Published in the United States of America by Cambridge University Press, New York

Cambridge University Press is part of the University of Cambridge.

It furthers the University's mission by disseminating knowledge in the pursuit of education, learning and research at the highest international levels of excellence.

www.cambridge.org
Information on this title: www.cambridge.org/9781107626454

First published 1933
First paperback edition 2013

A catalogue record for this publication is available from the British Library

ISBN 978-1-107-62645-4 Paperback

To

GEORGINA, NICHOLAS *&* JOANNA

May these Letters and Diaries help to preserve your father's memory, and give a closer and more personal picture of his life and character than can be found elsewhere

CONTENTS

PORTRAITS

PREFACE

A well-planned and well-written book can seldom need an introduction by another hand, but perhaps a book of memoirs may be one of the exceptions to this rule. It is natural to hope that a personal narrative may gain in vividness or in roundness if a touch or two can be added by a second observer, a friend who has stood outside the group of characters portrayed, but in continuous and intimate touch with some of them. It was this hope which moved me to accept the task when it was laid upon me.

My knowledge of Sandy Wollaston dates from long ago—he was the second son in my housemaster's family at Clifton, and was born about a year before I entered the School. My real memory of him may be said to begin in 1881, when I was leaving for Oxford and he was a fair intellectual boy of six, persevering at his violin and bright in conversation—though he was not so free of speech in the dining room, where I generally saw him, as he was on the staircase where he loved to lead a toboggan train, and in the upper part of the house, from which his voice would come ringing in masterful tones. His quiet studious side was I thought well represented by the portrait painted by Miss Mary Tothill, but my own imagination brings back to me an aspect which is even more characteristic—a large-eyed aquiline look and high forehead, as of a young eaglet barely fledged, but already bold and observant. Beyond this I remember little of him as a boy—I was going away into a wider world, and he was just 'one of the young Woolly Bears'.

But 'housefeeling' is in the most favourable cases a form of near kinship, and as time passed I found myself more, and not less, interested in the Wollaston family. During the next eight or ten years I was constantly revisiting Clifton, and whether it was in term or out of term the house in College

Road was still as in old days overflowing with music and conversation and hospitality. The Wollastons lived a full life, based on the service of a great school, but with windows looking out on a far wider horizon. Mrs Wollaston was a Richmond, an accomplished hostess, a pungently witty talker, the friend of many distinguished people and the critic of many more. George Wollaston himself was a Master on the Modern Side: in school—where I was never under him—he taught science and languages, but every day and at all hours in his own rooms he was giving us a course in history, literature, internationalism and the art of travel. By his drawings, his familiarity with languages, customs and scenery, his specimens of odd minerals, we got to know of his habit of solitary tramping in foreign parts and his delight in their inhabitants; but it was chiefly the birds, beasts and butterflies which attracted him, and the physical aspect of the countries in which they lived. I have more than once made an attempt to draw the outline of this singular and powerful figure in my school life, and I cannot here go further with it. My present point is to mark the nature of the inheritance which came by birth to his son. Sandy was in short the descendant on one side of a clan devoted to music and the arts, and on the other of a family which had been known in the world of science for some two hundred years. He was above all a born wanderer and explorer, and possessed many of the qualities which are invaluable for comradeship or for leadership in out-of-the-way places: two of these were his love of climbing, and his practical knowledge of medicine. This last gave him in a curious way both his first start in life and his second: he qualified as M.R.C.S. and L.R.C.P. and thereupon discovered that 'medical practice as a means of livelihood does not attract me—in fact I dislike it all extremely'. But no doubt he found himself doubly useful in his first collecting journeys—in the Sudan in 1901 and in the Far East in 1904—and so transformed his 'practice' into a career.

The pages of this book present his life as one of kaleidoscopic variety—there is simply no room in the volume for the number of maps which would be needed to illustrate his travels satisfactorily.

His friends rejoiced in his successes but to the nearest of them his repeated disappearances and long periods of absence were a continual regret. We were proud of him, and had an unlimited belief in his power of pulling through: but when he wrote or came home in person we felt as though he belonged to a fourth dimension which we could not know, even when he yielded to our urgent requests and wrote about his travels.

From the beginning he had a patient conviction that he could not write anything good enough for the reading of the public, and it pleases me to believe that that conviction was first undermined by the publication of his letter to me in 1906, describing Lake Naivasha, which he visited on the way up from Mombasa to Nairobi. It conveyed so convincingly the beauty of the place and his own peculiar delight in it, that I ventured to send it on my own account to my friend J. A. Spender, who at once printed it in the *Westminster Gazette*. I was amused long afterwards to find that Sandy had himself shared our opinion of its merits: he says on February 26, 1906 (page 85 of this book) that 'I wrote rather a jolly thing about Naivasha as I was walking along the road from Entebbe'. Then apparently it 'stuck fast' for a time, but in the end led him on again to the publication in 1908 of his first book *From Ruwenzori to the Congo*. This gave an excellent account of a modern scientific expedition, a group of comrades intent not only on collecting all possible evidence of the fauna and flora of a region hitherto unknown to the scientific world, but also bent upon the survey of mountains not yet accurately mapped or measured. The impression left upon the reader's mind is one of a fellowship working in conditions of great physical health and almost ceaseless effort and enjoyment. When I re-read it after the War I realised that those who go upon one of these scientific expeditions are in reality entering for an international competition of a new kind, one in which the competitors have none but peaceful and unacquisitive intentions: in which the success of one is in a sense the success of all, and the national rivalry, the desire to be first at the goal, calls forth a helpful rather than a hostile feeling. The story of the Ruwenzori climbing race between Sandy Wollaston and the distinguished Italian climber—the Duke of the Abruzzi—is a tale of chivalry and courtesy throughout: and though the peak named 'Wollaston' is a few feet lower than the Italian summit, it gives the loftier reason for national pride and remembrance, as those will realize who read our man's candid and sympathetic account of the race. Another example of this new kind of Olympic Games will be found in Wollaston's *Pygmies and Papuans* published four years later, after his return from his first visit to New Guinea. This book is, for me, even more interesting than the Ruwenzori volume—it touches the imagination as even a brilliant work of art could scarcely do. The Papuans when Wollaston visited them had no knowledge of the use of bronze or iron—they were at the stage when men had perfected the use of stone, wood, and bone for implements, when they had begun to cultivate crops for food, and to keep domestic animals. They belonged in fact to the Later Stone Age, the Age of Neolithic Man. Six or eight thousand years ago our own ancestors in Europe were Neolithic men: in looking therefore at the Papuans Sandy and his companions realized that they had outdone all their predecessors—they were looking at the daily life of their own race with all its common acts and contrivances in a period so remote that we have and can have no written record of it. They saw it not as a picture or a scene imagined, but as a real life however ancient—a life which they could share to-day, though separated from it by thousands of years of time. 'I remember', the record tells us, 'seeing him (the stone smith of the village) sitting outside his hut sharpening an axe, with three or four others lying beside him waiting to be done, while a few yards away a woman was splitting a log of wood with a stone axe. It struck me as being one of the most primitive scenes I had ever witnessed, really a glimpse of the Stone Age.'

This was the last travel-record which Sandy published, and it is a fine climax. But much work lay before him yet and the story of his second expedition to New Guinea may be read in his diaries included in this book. In the present volume will also be found his answer to many of the questions which his friends would have put to him about his experiences in the War—his long cruises in the *Mantua* and his view of Scapa Flow from the *Agincourt*, his voyage with General Sir Horace Smith-Dorrien to the Cape, his success in bringing away the sick and wounded from the Rufiji River, his over-time service in H.M.S. *Humber* at Archangel and in the Dwina River, where fighting went on for months after the Armistice in France; until we reach the entry on October 23, 1919: 'Demobilized to-day—Laus Deo'.

Then follows the Epilogue—the return from War—which to Sandy was always uncongenial—to the old joys of mountaineering, the Everest Expedition of 1921, and in 1923 the visit to the Sierra Nevada Mountains in Colombia. Last of all the home-coming to King's. 'It is a big adventure again, but what has my life been but one adventure after another? and I have been simply dogged by good fortune.'

HENRY NEWBOLT

MY thanks are due to many friends and relations who have kindly allowed me to read through and publish the letters they held from my husband. I am also indebted to those who helped me with the MS. of this book and gave me advice.

To the Royal Geographical Society I am grateful, for their permission to publish some remarks on New Guinea made by my husband at one of the Society's lectures.

MARY WOLLASTON

Cambridge 1933

I

CAMBRIDGE AND BIRDS

EARLY YEARS

Alexander Frederick Richmond Wollaston was the second son of George Hyde Wollaston whose forbears for some two centuries had been distinguished men of science. George Wollaston himself was a man of great culture and wide interests, and an outstanding figure for many years as a master at Clifton College. A.F.R.'s mother was a Richmond—with both her father and grandfather distinguished in the world of painting. She was a woman with an unfailing zest for life, delighting in people, art and literature. She was a great lover of music and a fine musician.

Most of A.F.R.'s boyhood was spent at Clifton, for after leaving his private school he returned home to enter Clifton as a day-boy. He was not the kind of boy to be interested in school life or games: his heart lay elsewhere, above all in the study of birds and the delights of the countryside. He was a born naturalist. He was still at his first school when he formed the habit of keeping 'note-books'. In these he noted down anything that struck him as being of beauty and interest, and although it was more especially of birds that he would write, there seemed little in nature too small to attract his attention and excite his wonder. I have, however, decided to begin with his Diary of May 1893, when he was nearly eighteen, and spending his last term at Clifton before going up to Cambridge.

MAY 5, 1893. (*Clifton.*) Heard first nightingale. Went to Sea Mills and across bridge and along railway for about half a mile. Here I watched a pair of whinchats and saw the hen go to the nest twice in about 50 yards; then I walked up and found the nest with six eggs. The birds were both about, but did not show much excitement.

MAY 13. To Leigh Woods for wood-wrens. The birds did not seem as excited as they would be if the hens were sitting; I expect they have not laid eggs yet or at any rate not the full number. I found a redstart's nest with six eggs, three of them very curiously longshaped. In the evening I tried to find out the locality of the nightingales, but the beasts would not sing. Heard several nightjars. (Some females I know think the latter are telegraph wires.)

MAY 20. On the screezy slope below the short downs, about half-way down amongst some ground ivy, I have just found a nightingale's nest with five young birds. This is the first I have ever found, and I did this by watching the birds go to their nest with grubs in their beaks. I could not make out for certain whether both birds make the plaintive whistling alarm note: the cock certainly did, and it is the cock also that makes the harsh 'krrr'.

MAY 31. Went to visit my nightingales. The old birds did not come near the nest while I was about and did not utter a sound; the young birds are pretty nearly fledged now.

JUNE. This month I found my first marsh-warbler's nest. It was not unlike a whitethroat's, suspended from a stem of meadowsweet and an osier stem. The birds were exceedingly interesting. The hen slipped quietly off the nest when she was approached, while the cock sat on the top of an osier giving a most amusing concert. He began by singing his own song —which was like a reed-warbler's only more so—and then went off into the songs of other birds. I heard distinctly the skylark, great tit, whitethroat (both song and alarm note), swallow, willow-wren (song and alarm note) and whinchat. Besides these this marsh-warbler imitated more or less well several other birds. It was interesting to compare him with the reed- and sedge-warblers which were numerous in the same osier bed, and there ought to be no great difficulty in future in distinguishing the marsh-warbler.

This summer A.F.R. left Clifton, and in October 1893 he went up to King's College, Cambridge. He writes to his sister from King's saying:

I like this place tremendously and I have now pretty well got into the ways of it. Well—to tell you what I do. I get up at any time between half-past seven and half-past nine: at least four times a week at half-past seven, so as to be in time to 'sign in' at the Porter's Lodge by eight o'clock: this is an alternative to going to chapel for a very dull service, and, as you may imagine, it is not very thickly attended. It is rather amusing to see men tearing to the Porter's Lodge at about one minute to eight, clad in not much more than a fig-leaf, sometimes a collar-stud or eye-glass. Then comes breakfast: that is ordered by leaving a note on your table for your 'bedder' to find when she comes in the morning. You get a splendid breakfast for sixpence and of course you can get much more if you pay for it. Then comes work: lectures three days a week till lunch, which is between twelve and two o'clock. After lunch some form of exercise—generally tubbing (rowing two in a boat). Then tea, and chapel at five: this latter is voluntary and I nearly always go. Hall at seven, and after Hall you go out to coffee and smoke in other men's rooms, or they come out to you. Here at King's the second and third year men don't call on the freshmen, but they ask them out to coffee after Hall. Then comes work, and after that bed—about twelve o'clock. Now I think you know pretty well what I do with myself.

A month or two later he writes again:

(*King's.*) Without exercise I am sure you would die here, for it is the most deadly climate I was ever in. If I stay in in the afternoon I am bound to go asleep, and then it is impossible to wake up for the rest of the day. What with the draughts and fogs and chilliness it is enough to make your blood run cold. However, apart from these little inconveniences it is a most beautiful place. The sunsets are magnificent and the trees have been lovely, but of course they are mostly bare now. There are many very nice men here that I am beginning to know, though it takes some time for me to get to know people. Your exhortation to me to work hard is fortunately needless, though it might have been of use at the time when I wrote to you before; for the day after that I more than doubled my work by adding two more courses of lectures which require a tremendous lot of extra reading, so now with three different courses of Geology lectures, besides Biology ones, I have every prospect of turning into a fossil before the end of this term. In fact I have got such a frightful amount of work to do to-night that I don't know what to do first, so I will think it over while I write to you. (*A fortnight later.*) Here I think I must have gone to sleep or done some work; I can't remember now what I did. Anyhow I lost this letter and it has only just turned up under my inkstand.

FEBRUARY 15, 1895. (*Cambridge.*) Great frost. Cam frozen.

FEBRUARY 25. Winter aconites flowering in King's avenue.

FEBRUARY 28. Snowdrops flowering behind John's.

MARCH 1. St David's Day. First day of spring—my annual holiday.* A chaffinch is singing in the trees behind Paradise. This is very late for them to begin: generally I hear them begin a feeble attempt at a song in February, and one year I heard a bird sing in January. Swallows are building near the town bathing sheds. Rooks busy building everywhere, lots of beetles on the wing, and at Byron's Pool—where I sat and basked in the sun—I heard a corn-bunting sing. I also heard a pied wagtail on the Backs; don't remember to have heard it before.

* A.F.R. had what he called a 'self-persuasion' that spring began on St David's Day, and whenever possible he would make that day a holiday for an excursion into the country.

A.F.R. IN 1894

MARCH 29. (*Back at Clifton.*) Wasps flying about everywhere. Chiff-chaffs migrating along hedgerows near Pilning, down the Severn from N.E. Where did they get into the Severn valley? They must have crossed midlands from East coast, and perhaps followed the Trent, Ouse, etc. Saw three wheat-ears on the Downs—very tame, evidently just arrived. Heard two cock robins having an exciting singing match on the downs; so jolly they were.

APRIL 4. (*Braunton.*) Waded in to the dipper's nest that I found two years ago under railway bridge. Eggs hatched already; they must have been laid by the middle of March. I saw from the train, when I was going to Teignmouth, a whimbrel on the Tor; also caught sight of a dipper that went plump feet first into a stream. Stayed at Bovey Tracey, and walked on to the top of Hey Tor rocks. Such splendid country—very wet and streamy. Numbers of cormorants on the Teign.

APRIL 23. (*Cambridge.*) Watched a common sandpiper on river between King's and Clare bridges. He flew up and down and settled several times, then flew up river to where? Nightingales very good this evening. The cock nightingale is a most beautiful bird—spread tail red.

APRIL 30. Went to Wicken Fen. Had a splendid view of two grasshopper warblers sitting on a low bush and singing. At first I thought the birds I saw were something else, for the song seemed to come from much further away. Saw a great many snipe, but I could not find their nests. Their drumming is most extraordinary and I am almost convinced it is made by tail feathers in descent. Watched through glasses and when the birds came down I saw tail feathers spread out and vibrating very hard almost immediately after; just time for sound to travel. The drumming continued until the bird began to ascend, then it immediately ceased. This means that, allowing for sound to travel, drumming ceased before bird began to ascend, which might naturally be expected as tail feathers would not vibrate very rapidly at bottom of curve of flight.

This evening I heard the first young rooks in nests in Backs.

MAY 2. Went into Botanical Gardens and saw a pair of nightingales hopping about. They were disturbed when I went near. Too many people about for me to search. Went into a strip of wood by the Trumpington Road and walked straight up to a nightingale's nest close to where I had found one last year. Later I returned to the Botanical Gardens and there found a nightingale's nest where I had marked a pair of birds. It was a very pretty nest in stump of elder.

JULY. This month I heard several curlews flying over at night—especially dark, cloudy nights. I suppose they come closer to the ground on such nights. On one clear night I think I heard a redshank flying over; he sounded a long way off and must have been at a very great height.

One day I walked along the coast from Hunstanton and found the nests of two lesser tern. The eggs were laid right down on the shingle not far from high-water mark; no attempt at a nest was made, and so I think it was a good deal harder to find than a ringed plover's. On my way home I several times heard a loud squeaking, more like rabbits in distress than anything else. I could not make out what it was, until a pair of water-rails got up at my feet making the same noise. I think this is the first time I have seen a water-rail fly.

SEPTEMBER. Went to stay in Arran. Saw heaps of birds but the only new species was gannet. They shine in the sun like snow—I never saw birds so intensely white. It is a grand sight to see them suddenly dive down beak first into the sea from a great height; tremendous splash. The black primaries are very conspicuous. All the birds I saw were adult.

JANUARY 15, 1896. (*Cambridge.*) Heard a chaffinch beginning to sing near Newnham; he tried the beginning of the song and failed miserably. Saw a kingfisher sitting on the willow just above King's bridge. Winter aconites flowering in King's avenue and the Wilderness of John's.

JANUARY 25. A thrush has been singing magnificently in the Provost's garden for more than a week now; every morning for two or three hours and sometimes in the afternoon. I went this evening to see Professor Newton* and we talked birds all the time. He told me an interesting thing about the bills of birds. In some, specially finches, the length of the bill in spring is so different from the length in autumn that the same bird has been described at different times as two distinct species. The beak grows enormously in summer when the birds feed on soft food, and gets worn down in the winter by seeds.

FEBRUARY 5. The young leaves on willow by King's bridge are quite an inch long. Trees are budding everywhere, and lots of thrushes are courting each other in the Botanical Gardens. One of my emperor moths that I got in Arran came out this morning; it is a female, nearly a perfect specimen. The warmth of the room must have forced it; it had been rattling about in the cocoon a great deal lately.

FEBRUARY 26. Saw a flock of peewits flying northwards over the Backs this afternoon.... Sallows budding by bathing sheds.

MARCH 20. Chaffinches improving steadily. Plums beginning to flower, and the quince tree in New Court getting green. I was woken up this morning by a chiff-chaff singing in the Provost's garden. Went for a long bicycle ride and was delighted to find a cowslip out in a field near Huntingdon.

* In 1921, A.F.R. wrote the *Life of Alfred Newton.*

APRIL 4. (*Braunton.*) Went home and came to Braunton. Looked at my little dipper's nest under the same bridge where I have found it in '93 and '95. There are two young birds about half grown in it; eggs must have been laid very early. I found a linnet's nest ready for eggs in a gorse bush, and a kestrel's nest ready for eggs in a fir wood up the Chattowell. A tremendous lot of *Equisetums* all over the water-meadows.

APRIL 15. (*Porlock Weir.*) Went to a meet of the D. and S. staghounds at Hawcombe Head. Two hinds broke away from Birch-hanger plantation very soon after tufters were put in, and went away towards Black Barrow. We had a grand view of them trotting down the coombe and up the other side, followed by a few hounds. Then appeared three more hinds, and after that general confusion for the rest of the day—deer and hounds all over the place. Coming home I nearly trod on a very fine adder near the cross roads below Hawcombe Head. It was the best marked specimen I ever saw, very silvery and deep black markings. We stunned it and then got it into a field-glass case and afterwards took it to the Clifton Zoo. There are a great many ring-ouzels about on the move. One puzzled me much by making a very curious plaintive whistle like a bullfinch, but really quite distinguishable.

APRIL 24. (*Cambridge.*) The double white cherry in Fellows' garden is now in flower; a most beautiful sight. I heard a lesser whitethroat in Trinity garden and saw a spotted fly-catcher in John's Backs. I also heard and saw a wood-wren in the lime trees of King's avenue; this is the first I have ever seen near Cambridge.

MAY 2. Found a lesser redpoll's nest ready for eggs. The birds made a great noise when I went near, and they fetched another pair and all began to make a great fuss; there must certainly have been a second nest nearby.*

MAY 3. In Wicken Fen this afternoon I saw a large hawk which I immediately identified as a harrier. Rather light grey, wings long and pointed, rump very pale. It was probably Montagu's harrier which is said to have nested there years ago. It was a glorious afternoon, and the bees and smell of thyme wonderful. In the evening I went up the river to see a goldfinch's nest that I found a few days ago, but the whole thing was destroyed; damnation.

MAY 11. To Wicken Fen to search for snipe's nest, and just as I was coming away I had a splendid view of a short-eared owl—the first I have ever seen. He sat in a birch tree within a few yards and looked at me. I went a few days after to search for his nest, but failed to see even the bird again. (Heard afterwards that they safely hatched eggs.)

MAY 13. Basked in the sun nearly all to-day and spent the evening in the Fellows' garden with Richmond. We heard a goldcrest. This morning in the Backs I heard the curious roaring noise made by stock doves. There was a note about it in last week's *Field*, otherwise I should not have recognized it.

MAY 17. Went with Gayner to Lakenheath. We walked towards Thetford through very jolly country, full of marshy patches and sandhills. Then got to a deserted place called Wangford where we drank much water. On for about half a mile through warren and pines without seeing any birds and then I saw a stone-curlew flying off in the distance. Searched a small patch of flints lying very thick on the ground and found the nest with two eggs. (Result of Seebohm's tip of looking for two flints lying together.) A good number of

* In May 1899 he again notes this redpoll's nest and says: 'This is the fifth year that I have known them to build in the same clump of honeysuckle in the Botanical Gardens—always within a few yards of the same place'.

rabbits' droppings had been scraped together to form the nest. Went on for about two miles till we came to a long wide waste patch covered with flints and rabbit burrows. We quartered the ground carefully and found another pair of eggs....Stone-curlew make a very loud whistling quite unmistakable. It was disappointing not getting a better view of the birds themselves, but finding their nests made up for it to a great extent. This is quite my best find next to the marsh-warbler. I must go again—the country itself is quite worth going for.

MAY 18. Went again with Gayner, and outside a wood near Trumpington—on the sheltered side—we saw hundreds of swifts hawking flies close to the ground and very slowly. Gayner made a hit at one with his umbrella and knocked it down dead; soon afterwards he did the same with another. It was horrid murder, but very remarkable that such a rapidly flying bird should be knocked down so easily. We visited the nest of a whitethroat, and I was astonished to find six pure white eggs in it.

MAY 20–JUNE 16. Bicycled through Harston one day where the laburnums are very beautiful and the smell of a bean field delicious. Am having a poor time generally with Tripos at intervals; most annoying when the weather is so beautiful.

JUNE 23. Took degree and came down from Cambridge.

Some years later, when writing in sympathy to a sister who was in the throes of her Tripos examination, he says:

I remember waking up to the fact that it was too late to begin to work, so I settled down into a sort of philosophic calm and enjoyed the term almost as much as any other, except during the days of the exam when the weather was very hot and fine and I would fain have been elsewhere. Fortunately 'it will all be the same a hundred years hence' whatever people may say. I know well that the friends you make at

Cambridge are worth all the Firsts and all the triposes that ever tripped—or I was particularly blessed in that respect.

After taking his degree this summer A.F.R. stayed up at King's another two years, finally leaving in 1898. No Kingsman could have loved his College more than he did. He said once in a letter to me: 'I know that I have loved the College and Chapel more than any other spot in the world. From the beginning King's became a new home for me, and I have been more proud of being a member of King's College than of anything else in my life'. It is characteristic of him that in 1906, when he lay sick in a Congo camp, he should note in his diary: 'December 6, St Nicholas and our pious Founder King Henry VI. Thought much of King's and the peaceful beauty there. Drank "in piam memoriam" in a liqueur glass of Congo port'. Again in 1912, when cutting his way through the forests of New Guinea, he sends a line to a Kingsman saying: 'December 6, Give my love to King's. It's an odd thing but King's is the place where I feel myself more firmly rooted than anywhere else'.

II

TWO LAPLAND DIARIES

These two diaries were kept by A.F.R. in the summers of 1896 and 1897, when he went to Lapland with his friend Herbert Playne.

I

AUGUST 2, 1896. Somewhere between Molde and Christiansund we saw several Richardson's skuas which I first mistook for falcons. There were two forms of them—one dark all over, and one with white belly and sides of neck. They flew at a great pace, making straight for a gull and tormenting it till it dropped its food. The long middle tail feathers were very distinct.... Proceeded to Trondhjem and arrived at midnight when it was still quite light. The captain of the ship said that the long pale streaks which were spreading over the fjord were made of herring spawn.... Left by train for Storlien. Saw a young cuckoo in charge of a willow-wren, and the first greenshank I have ever seen: its shape is much like that of a redshank and its legs conspicuously green: the wings are entirely brown—not with white secondaries as in the redshank—and the bill is slightly turned upwards at the tip. Travelled all night right into Sweden, and then through rather monotonous country; hundreds of lakes of various sizes and a few grand rivers with masses of logs floating on them. Got to Lulea in the evening and found that the boat to Haparanda takes twelve hours and not two as we expected. The place is like any ordinary German town: straight streets, square wooden houses and a hideous church. Walked up on to a hill overlooking the harbour. The sea looked like a great lake covered all over with heaps of flat islands, black with fir trees. The nights are so light that as we were leaving the harbour I could plainly see the time on the church clock, more than a quarter of a mile distant. I heard a band in the town playing 'After the Ball is over'.

AUGUST 6. At Tornea we went to the inn and had a real Finnish meal, beginning with all manner of little things—lax and reindeer and a horrid kind of drink. Then we strolled about in the afternoon and slept behind some piles of hay. We met a man who talked English. He told us that the river had once flowed on the other side of Tornea and this accounts for the town being always marked on the east side of the river.... Left by cart and travelled some thirty miles to Piipola. Stayed at a house full of men, women and children of all ages, all most remarkably clean. My bedroom is between the hall and the part of the house where the whole household sleeps. They all troop through making a hideous noise up to midnight, and begin to go back again at five in the morning....

Walked down to some rapids this afternoon and had a very jolly bathe, which was a bath as well, as I had not had one since I left England. Had an excellent view of an osprey fishing. He went down with a great splash into the water, came up, and struggled on the top for some time with a fish, and then flew off holding it in his claws so that it hung directly under him—the head pointing forwards. There is a great salmon weir across the rapids with nets at intervals; the natives stand on little projecting platforms and scoop out with their nets hundreds of small salmon which lie waiting to go up the river; the fish weigh mostly from two to four pounds. Saw a few black-throated divers and an old hen red-breasted merganser with a brood of young ones.

AUGUST 9. Forgot to look out for the eclipse though we were both awake at the time.... Left here and travelled on by cart to Tervola and then by two more stages to Takkunen where we spent the night. Roads fearfully bad. Crossed the river twice by ferries, and the small tributaries by bridges with very steep slopes on each side—rather like a switchback. Stayed at a clean little inn, and there was no nocturnal procession of inhabitants through my bedroom as I had had at Piipola.

AUGUST 10–25. Drove 25 miles to Rovaniemi. Had a very good view of a pair of great grey shrikes. Along the river valleys are long low hills, none more than about two hundred feet high, but enough to make a great difference to the beauty of the country. There was a glorious sun this evening and the colours and shadows on the hills were most beautiful. Hitherto the landscape has been rather featureless, but this has been made up for by the wonderful colours of the forest and water in the early mornings and evenings.

We rowed up the eastern branch of the river to-day, and about a mile up we met a crowd of men in boats pushing on the stranded logs that float down the river. They make a great rope of logs tied end to end and stretching right across the river about half a mile broad; when the ends are pulled the logs are swept along in front leaving the river free above. It was a fine sight—the men in their big canoe-shaped boats tied head to tail, dressed in bright coloured shirts, blue and red, smoking cigars and singing a strange sort of chant. We then walked about on a sandy sort of island, covered with a thick growth of *Equisetums*, growing about a foot and a half high. We put up a bird which I at once said was a little stint. Tracked it to the shore, had a long very good look at it through my glasses, and found it was different to the little stints that I had seen before. When we got back we identified it as Temminck's stint—a rarer bird. It is easily distinguished by having the outer tail feathers white; it is also more mottled on the back, many of the feathers being dark with a buff margin. It was a young bird—some of the down still clinging to the back of the neck and the base of the bill. I wonder where they nest. We also saw some great black spotted woodpeckers in some pine trees by the river. They made an unmistakable woodpecker cackle, also a very curious bleating cry—rather like an owl or hawk.

Playne and I have been strolling about watching birds, and fishing. We caught a small pike, touched one or two grayling but did not hook them; however, it was glorious to lie in the bow of the boat basking in the sun.... Heard a great quacking of ducks, and saw a pair of pintails flying up the river: tremendous big birds with long tails.... Watched a pair of black-throated divers. There is a very noticeable difference between the ways in which ducks and divers go under the water: the ducks seem to spring partly out of the water and go over in a curve like a porpoise, the divers gently sink down under the water and appear to be pulling themselves down by means of their feet from below.... Found out this evening from *Backhouse's Handbook of European Birds* that some of the birds which we have taken to be marsh-tits are probably small Lapp-tits: a very similar bird but having a brown instead of a black crown and being buffish on the sides of the breast and belly.

Playne fished this afternoon whilst I lay on the bank by the rapids and read some Milton. Saw a very small long-winged falcon—undoubtedly a merlin; first I have seen.

Caught a grayling about 4 inches long: this is the first fish I have ever caught with a fly! I find great difficulty in getting the line out straight....

Wandered into the forest and saw the Lapp-tits flying about with the marsh-tits. Am quite sure about the Lapp-tits, for Backhouse says: 'common in Lapland; birch and pine forest'.

I saw a little bird—evidently a young bird—flying into some bushes by the river, with many of the head and neck feathers tipped with buff: the tail dark brown at the end and the coverts reddish. The build and general appearance of the bird were very robin-like, and the attempt at a song was like that of a robin. Decided that it was a young blue-throat—the red spotted form which is found very far north.

We got up early and left Rovaniemi by the Kittila road, and walked thirteen miles of our way to Tornea. Had a splendid view of a golden eagle being chased by a pair of buzzards; he looked immense by the side of the buzzards and it was his great deep croak, rather like a swan, that attracted our attention first....Went westwards through rough country and swamps until we came to a small farm—kind of hay chalet—and here we slept; we were given coffee and very good sour milk. Continued the next day through fearful swamps, and saw many reindeer which seemed as much interested in us as we were in them. In the evening found another hay chalet on the edge of a swampy lake. We were very thirsty, but as the lake had no outlet we did not like to drink. Had a good long rest, though not much sleep, as the mosquitoes and midges came in swarms and nearly drove us wild. Saw many capercailzie and wild geese....Got up early next morning—fearfully bitten and very damp. More tramping through swamps and thick moss up to our knees. This kind of thing lasted for three or four hours but eventually we got on to a hill where we had a magnificent view of lakes, and what was better still, a house about a mile and a half away. We went to it as quickly as we could move, told the woman there to get us 'Ruaka' or breakfast, and then went down to the lake where we had quite the best bathe I've ever had in about 6 inches of water. Afterwards devoured large quantities of rye-bread and butter and a bowl of thick sour milk....

On along the track which we thought would lead to Alkula, but after much trudging it took us into an impassable swamp. Tried another path with the same result. Returned to the small house and lodged there for the night....Saw a pair of Siberian jays in a very thick jungle; also a greenshank.

Started off in a boat to the end of the lake (Vietonen), crossed a neck of land, and got on to Lake Miekojarvi; rowed down this lake for about 10 miles, then we went down a river until we came to a likely path, but after two hours' hard walking we found ourselves again in a thick swamp. Had to come all the way back and slept in a hay chalet. The next day was pouring rain. A man showed us the right path to Alkula, and after a tiring time through swamps and over lakes and losing our way many times, we found two men who escorted us along intricate forest paths which we would never have found ourselves but which proved to be a short cut to our nearest stopping place. Had a meal at a farm: the inevitable milk, butter and hard rye-bread....Heard several birds in the pine trees, but at first couldn't make out what they could be. They turned out to be waxwings. They seemed very tame, keeping in a flock, sitting on the tops of the pine trees and taking short flights from one tree to the other. They made a curious bubbling whistle quite unlike any note which I have heard before, so there should be no difficulty in recognizing them again.

Ploughed on through swamp, forest and cattle tracks, till we reached a lake the shores of which were made of fine granite sand, and the water the clearest I have ever seen. By the river Trugeli we found some log-cutters taking logs down the stream. We got into a boat with about fifteen of them and they rowed us across, singing a very curious song —sort of dirge—mournful, and with very long verses. After bread and butter at a small house, we went straight as a die over every conceivable sort of bad ground till we came out into open valley and on to a road which meant the end of difficult walking. Here we were told that Alkula lay 4 kilometres away. We walked about five and found we had ten more to go. It was a great blow as we were very tired and hungry. However these 10 kilometres went faster than they might have done owing to the sight of a family of willow-grouse, new birds to both of us and so tame we might have touched them with a stick. At Alkula we hoped to get a meat meal but were disappointed. With difficulty we got eggs to add to our milk and rye-bread. Went to bed very tired and rather cross. This inn was so uncomfortable that we decided to push on a stage towards Tornea. We walked 14 kilometres to an inn where the bread was quite uneatable, so drove on another stage to a place where the food was almost equally bad although we liked the coffee made with salt,—a common custom in these parts. In the evening we lay in an old boat by the Tornea: a grand stream and it goes at a grand pace. Next morning we again walked many kilometres hoping to find decent food at the end, but again we failed: not even an egg,—just rye-bread and butter. We imagined we still had three more stages into Tornea, so we hired a cart for the first stage, intending to walk the other two, but to our great joy we found that the end of this stage landed us only about 3 kilometres from Tornea. At Tornea we walked over the bridge into Sweden and had an excellent cup of tea,—the first since leaving home.

In a letter to a friend telling of this journey into Lapland, A.F.R. says:

There were no roads of any description and hardly any tracks, as the natives, who are very few, live by the lakes and do most of their travelling by water. At one time we went along for more than twenty-four hours without seeing a sign of a human being. Our greatest difficulty was the swamps, which often stretched for miles and had to be crossed somehow. It was fearfully hard work jumping from one safe patch to another, or wading knee deep in mud or wet moss. And the mosquitoes!...At the end of a week I was covered with grime from head to foot, my shirt was black, the feet of my stockings had gone long ago, and the various parts of my boots were tied together with string. I have a hazy idea now of what a tropical forest must be like. There are no hills more than 500 or 600 feet high, so you might imagine that it was a dull and featureless land, but it is really very beautiful —perhaps from the vastness of everything. When you stand on the top of a hill you see enormous stretches of forests and lakes and rivers; lovely colours, and such a clear blue sky, not like the deep Italian sky, but one that you can see through to the beyond. And the lakes and rivers. You know the colour of the Cornish sea on a blazing August day—well, they were that only more so....

II

AUGUST 1–8, 1897. Playne and I spent the day at Bergen in the Natural History Museum trying to identify a few birds. In the evening we loafed about the town where most of the inhabitants seemed to be doing the same. On August 2 we crossed the Arctic Circle about 8 a.m. Coast very fine, mountains pointed and twisted in all manner of queer shapes. Saw the Loffoten Islands some 48 miles from the mainland. The ship ploughed her way through twisting channels between thousands of big and little islands. I'm told the pilots steer by watch and compass in the winter as there are no lights to mark the channels; rather close work, as some of them are not more than 200 yards wide.

We arrived at Hammerfest at midnight. All the inhabitants seemed to be out of doors: women washing and men painting their houses—in fact everything as if it were midday. Saw the captain of the *Fram*, Otto Svedrup, just going on board his ship; had a long look at him. He starts for Spitzbergen this morning. Lucky dog, wish I were going too. Up the Alten Fjord to Bugten, which is our nearest landing place to the mouth of the Alten river, and here we started off on our long walk. Spent the night on some hay in a hut. Saw some willow-grouse and I managed to shoot one, so perhaps we shall not starve after all. Met someone who said we couldn't possibly have a gun without a licence, but there doesn't seem to be anyone likely to stop us so we'll risk it. Saw some birds which puzzled me; they are very much like redshank, but have no white secondaries.

Left the fir-tree country behind us and mounted a broad plateau where we saw several dotterel. Playne shot one which we ate and found excellent. We also shot two mallard and Playne caught several trout—one of them weighing about 3 lb.; enjoyed them much for supper. The Lapp women are very interested in our clothes and possessions, especially boots; they dress entirely in reindeer skin, with a tight skull-cap of red and blue wool. Slept on and under reindeer skins.

Saw some grey plover in a marshy bit by the road: shrill plaintive whistle theirs is. I caught sight of a shore-lark and a dotterel with two quite young birds in down; they were very yellow and spotted with black.... The rain and mosquitoes are terribly tiresome.

Up a fairly steep hill to-day and lost our track. Had to ford a big river before we got to Kilpis-Järvi. The people here are entirely Lapp—only talking a word or two of Norse. After supper an old man Lapp and his son came in to pay us a state visit. We smiled and grunted at each other for about ten minutes and then they departed.

AUGUST 9–10. Got up very early to-day as we had a long journey before us. Began by losing our track, picked it up again, then lost it entirely. Eventually, after struggling through swamps, we hit upon the Alten river far below our proper point. We had walked for ten hours and so were tired and very hungry when we got to a hut, and there had a splendid meal of boiled peas, bread, and milk.

We decided to go by boat to Kontsheino. The boat was

very leaky and small and rowed by an ancient Lapp who seemed to enjoy the torrents of rain that fell upon us all. At Kontsheino this old Lapp took us to a house where we understood we should be given food, but after waiting half an hour, getting more and more hungry, he led us away to a house where we sat in a gorgeous sitting-room with a lady, who exchanged greetings and grunts with us till things began to get so strained that we got up to go; fortunately she understood in time that it was food we wanted, and at once prepared an excellent meal of beefsteak, cloudberries and cream, for all of which she would take no payment. We thanked her very much and came away. We determined to get on to Karasuando, our nearest point on the Muonio River, and after about four hours' walking completely out of our way, we found ourselves at Sieppa, where there were two houses inhabited by Lapps. We went into the least unclean, asked for food, and got them to let us sleep in one of their stone houses outside. Had an excellent night on skins. The next morning we were guided to a small place called Oscar Järvi, and from there we went on to Aiti-järvi. There the woman in the house was cleaning wool fresh from the sheep's back with curious wire brushes; she rolled it up into long light rolls which she then spun with a spinning-wheel; it was a pretty sight and neatly done.

AUGUST 11. Left in the afternoon for Palojärvi, which is over the border in Finland. We kept upon a high ridge with magnificent views north into Norway and south into Finland. Playne shot a rypa. We ran into a Lapp encampment with a large herd of reindeer penned off on to a little promontory in a lake; several hundreds of them. It was interesting to see a Lapp pick out a hind from the herd, throw a rope round its horns, and hold tight while another milked it. We borrowed a pot from the Lapps, made a fire, and cooked a duck which we had shot in the morning; then ate it surrounded by a crowd of admiring Lapps. We amused each other immensely and I was sorry to go. The camp was exactly on the frontier between two lakes about a hundred yards apart: the water from one going down to Tornea, the other north to Alten. Got to Palojärvi after a fourteen-hour tramp. Here the Lapps were rather scared of us. They came out with a gun, but after a time we persuaded them to let us sleep in one of their stone houses.

Walked to Leppa Järvi, where we got bread and butter and milk which we could have enjoyed but for some putrid fish which the people put on the table before us....On for about an hour, lit a fire and cooked the rypa Playne shot yesterday; then on to Soantojärvi. Fir trees began to show again this morning (Lat. 68° 35′); ling began at the same time. Since we crossed the frontier we have left everything Lapp behind us: the houses are different, the people are different, and their clothes the ordinary Finnish costume. They seem very poor, and there is great difficulty in getting food or a place to sleep in; but we both slept well on hay and woke to a glorious day.

Walked all the morning and halted for about an hour to cook a teal on spits at a fire. Had a dip in the river while the bird was cooling. I'm afraid it had very little on it and we rose rather hungry. Got a good deal off our track and did not reach the Muonio River till the evening—both of us ravenous. Made our meal off coffee, rye-bread and butter; slept between blankets for the first time in ten days.

AUGUST 13. Up at six o'clock and along a good track which we hoped to keep but lost completely in half an hour. Made a long circuit to the east round a big swamp, hoping to pick it up on high ground, but failed to find it. These swamps are fearful, worse even than those we had last year. To a large extent I was consoled by seeing a pair of cranes flying round and making a tremendous 'tonking' noise. It was just about here that Wolley found them breeding for the first time forty years ago. Saw some Siberian jays. After eating our last crust of bread we decided to make for the river, got there eventually, and followed it down for some time hoping to find a house. We sighted haymakers on the Swedish side and hailed them. They brought us across the river and directed us to a house where we had bread and milk. The house was filthy and smelt horrible: very different from all the Finnish and Lapp houses that we have been in. It is curious that crossing a river should make such a difference. In the evening we found a man to take us to Katkesuando, about 7 miles on the Finland side.

To-day a red-bearded Finn insisted on guiding us to a place called Puranen—more than half-way to Muonioniska. It poured with rain the whole time and we were drenched through. Our path lay along the banks of the river and through hay swamps, but we got to Muonioniska before evening. It seemed a large city to us, though there are probably not more than thirty houses and one hideous church. Our inn was fairly comfortable.

Slacked all to-day and enjoyed it much, as we had walked continuously ever since landing. The inn people give us whitish bread, rather like teacake, which is a great change from the jaw-breaking rye stuff we have been having and shall have until we get to Haparanda. We had a sumptuous midday meal off two perch, one small pike (all with their scales on) and potatoes.

Went 20 miles along the Kittila road, which soon degenerated into a very rough track bringing us to a place called Roochacha; very jolly by the lake. The road we had followed puzzled us a good deal: in some places it almost disappeared, yet in others it seemed quite recently mended and new bridges made. It is marked in the Russian ordnance map of 1877, so it is possibly an old road fallen into disuse.

Did another 20 miles. Road very much the same as yesterday—well marked but quite unused for the most part, with new bridges over the wet places. Wish we could find more to shoot, for we have had no meat for some days. Saw a large family of capercailzie and of rypa but could not get near enough for a shot. Had a splendid view of a three-toed woodpecker—very much like a great spotted, but had a yellow crown and no red about it. Crossed our last watershed between the Muonio River and the Ounas. We shall now follow the Ounas until we are within a few miles of Tornea. Slept at Sirka-järvi.

AUGUST 18. Twelve miles along the same sort of road as before till we reached Kittila. This is the beginning of the 200 miles of road down to Tornea. Kittila is a large town of forty or fifty houses, and boasts of two shops and a post office.

Left Kittila and decided to drive the first stage and walk the next. The pony was the same sort of beast that we had last year—average pace 8 kilometres an hour: gentle amble along the level and slow walk up the least incline. Rainstorms on and off the whole day.... We left our pony in the afternoon and crossed over to the left bank of the Ounas, walking 17 kilometres to Waara. There we sat in a hut and I spent the evening reading John Milton.... Next day another 12 kilometres, and then we hired a cart which took us to Mertola.

To-day we walked and drove some 50 miles, the last part of which was over ground we had been last year on our way through the forest to Alkula. We arrived at Rovaniemi late in the evening, ravenously hungry. Made a tremendous meal at the inn with a dozen men who were having what they called a picnic to celebrate the new line of telegraph which was finished yesterday. The picnic was what would be called a 'drunk' in England: it lasted till five in the morning. Landlord Kouri was very hospitable and glad to see us again; we drank good wine with him. Basked in the sun on the top of a hill, as we had done on the Sunday that we were here last year; after dinner went to call on Herr B— who would scarcely believe that we had come across from the north without a guide. There was a Finn who spoke English well, and he told me that he had once been down from Kittila to Kemi on a raft in twenty-eight hours (about 190 miles); rather quicker than our rate of going. Rafts are not allowed to go down late in the summer because they destroy the fishing weirs.

AUGUST 23. We trolled in a boat this morning but caught very little. These last two evenings we have had a moon which did not set here at all; it is in its last quarter. We saw it high in the heavens all day long, and at about ten or eleven at night it was quite high above the horizon to the north.

Started for Tornea. The road was fearfully bad all the way, deep in mud due to the rains we have had, but our tramp over the rough country in the north has put us in splendid condition and we feel quite fresh after 18 miles along a bad road. Stopped for food at an inn where the landlady remembered us from last year and was very attentive with innumerable cups of coffee.

Had a good view of a flock of siskins. Newton and other people would not believe last year that we saw them as far north as this. There were swarms of Camberwell Beauties all along the road, and I caught half a dozen. At Tervola Playne went out in a boat to fish and I sat by the river and watched some men sorting logs that had got jammed on the rocks. These fellows run about on floating logs with extraordinary nimbleness and never lose their balance. They don't seem to work hard, and most of their energy goes in singing songs and in keeping up a curious prolonged shout all in chorus together: not exactly musical, but it is jolly to hear them, and they make a good sight with their long poles, top boots and scarlet shirts.

Another cloudless day. Left Tervola, and wasted about half an hour at a ferry, because the logs came down the river so quickly and thickly that we could not get across. Had a good view of cranes flying southwards, probably migrating; also a flock of geese going same way. From Piipola we walked and drove till we got to Tornea. A fair-sized viper crossed the road; it is the first snake we have seen in this country either this year or last....I saw a large hawk which I think must have been a goshawk.

AUGUST 28. Third class in Sweden is by no means comfortable. I spend most of the day standing on the platform outside the end of the carriage, talking to a Swede in German. Good for my German, but not very lively for the Swede. The line goes up some tremendously steep hills and we are sometimes more than a thousand feet above the sea. The country is pretty but monotonous: long stretches without a sign of a house or cultivation.

Rather boring as I have nothing to read and the train jolts too much to write. The line down this valley from Storlien to Trondhjem is very beautiful; it drops almost 2000 feet in the first 20 miles. Along the fjord we saw a great number of eider-duck, scoters, and other duck which we could not identify.

AUGUST 30. (*Trondhjem.*) I went to a barber who removed my five weeks' beard, and then I walked to the Cathedral, which is being restored. The choir and the transept are finished, but the nave is in a fearful state of dilapidation. Some curious old monuments and gargoyles, but for the most part I don't care for the building. It seems to be a jumble of a great many different styles of architecture. Some of the restoration is badly done, but the carving on the capitals of the pillars and the mouldings of the doorways are too elaborate and beyond the powers of the workmen. In the graveyard outside, each grave is enclosed by a railing, inside of which is a seat, and on every seat sits an old woman cheerfully occupied in contemplating the grave.

Sailed this evening. There is a small band on board which pays for its passage by playing during meal times and when we come into and out of ports. Coming into Christiansund this morning—through a very narrow rocky channel—it started so many echoes that it completely lost itself, and each man played just what he thought fit; the effect was very curious and quite as good to listen to as when they were all together.

At Bergen we changed on to a most palatial boat with all manner of luxuries, but unfortunately the sea was so bad that I could do nothing but lie in my bunk. Tried to have supper in the evening, but it was no use; if I could be sick I should probably be all right.

SEPTEMBER 3. (*Newcastle.*) Spent the afternoon in Hancock's Nat. Hist. Museum, where the collection of birds is most interesting, especially the case of Greenland and Iceland falcons. There is a beautiful collection of prints and drawings of Thomas Bewick—the originals of the head and tailpieces of his books—and a great many drawings of birds and parts of birds. I did not enjoy all this as much as I might have done because the whole place seemed to be swinging up and down from side to side like the confounded steamer.

III

BIRD-HUNTING AND CLIMBING

Having made up his mind that exploration was what he wanted most to do, A.F.R. decided on the medical profession as a stepping-stone to suit his purpose. He knew that as a doctor and naturalist he stood a good chance of being taken upon expeditions, but he could never have known how much he was to dislike medical study and medical practice.

In September 1898 he started work at the London Hospital.

OCTOBER 1898. (*London.*) Spent a Saturday to Monday at King's; ate lotus. Very jolly to see so many nice people again; they seemed to me to be even nicer than they ever have been before....It would be almost intolerable if I did not get away sometimes, to the Saturday 'Pops' and such good things; the Zoo would be a blessing too if I could get there more often, but it is generally dark before I can go.

NOVEMBER 30. Saw a kingfisher flying about the Serpentine.

MARCH 1, 1899. (*Bonner Road, Bethnal Green.*) 'First day of spring', but not much sign of it in this beastly place. Hospital all day; fog just thin enough to see that the sun is shining brightly everywhere except in this cursed place. Ye Gods, how I hate London!

MARCH 30. I took train to Dulverton, where I put on my rucksack for the first time after many months. Walked up the Barle to Higher Coombe and then out on to the open moor over Winsford Hill; turned at Chibbet corner and went to Exford where I spent the night. Next morning I went up through thick mist to Simonsbath, then up the hill to the head of the Exe and on into the good old country. There was a thick mist all the way until I got off the moor within a mile of Brendon—then up and down to East and West Lynn, under George Newnes' beastly new railway, and up a steep hill over a bit of outlying moor to Martinhoe; there I took a very steep short-cut down-hill to the Hunter's Inn at Heddensmouth. In the evening I read *Lavengro*. Next morning I went along the slopes of the hills above the sea to the Hangman, which I skirted, and came down to Coombe Martin; then along the coast to Watermouth and Ilfracombe; good sea views all the way....Next day to Braunton. It was a grand morning so I went over the Burrows to Crow Rock, where I hailed a man in a boat who sailed me over to the point of Northam Burrows. Then I walked by Pebble Ridge to Westward Ho! following the coast for a few miles, and along to Hobby Gate and Clovelly. A gorgeous day from beginning to end. Sat out on the quay all the evening and talked Navy and Kipling to a coastguard....Continued reading *Lavengro*.

APRIL. (*Hartland Quay*.) Stayed a fortnight at the Hartland Quay Hotel and did a considerable amount of work, but on the whole the weather has not been good enough for lying out on the cliffs as I should have liked. One day I walked all along the cliffs to Morwenstow and finding no one at the Rectory I took tea in the 'Bush' kitchen and heard all manner of Morwenstow gossip. Had not time to go over to Tonnacombe so walked back through Welcombe. Searched for buzzards' nests in the woods but without success. These coombes are very lovely in the evening light.

The landlord of this hotel, T. Oatway, is a very remarkable man and we made very good friends. He is the best-read man of his kind that I have come across: quotes Shakespeare, Longfellow, Ruskin and Swift, and he knows Kingsley by heart. This is a place to come to every year. Never shall I forget the great gale we had soon after I came. The sea was terrific and several ships foundered and were wrecked along the coast, and since then piles of wreckage have been washed ashore at every tide—masts, sails, tubs, barrels of lard and oil, ships' boats and lifebuoys. Really, it almost makes one hate this cruel sea in spite of its beauty; yet to-day it is a serene blue and as beautiful as ever—not a sign of the storm devil lurking below.

APRIL 19. Walked from Pilning to Aust Cliff and back. Standing on the cliff and looking up the Severn reminded me very forcibly of some of the views we got in Finland on the Tornea river: the general shape of the country—long low rounded hills (they ought to be covered with pines and birches)—wide flat grassy meadows (they ought to be swamps) sloping down to the broad swift river with here and there a low cliff—all this reminded me of many places I saw in '96 and '97.

MAY. Went up to Cambridge to work at Physiology, and spent a lot of time looking for birds. One day I went to Wicken to search for the short-eared owl's nest. As soon as I got near the Wicken side of the fen the male bird came flapping about my head uttering a short bark; sometimes flying some distance off and sitting on a reed or small tree, but invariably coming back and showing unmistakably that the nest was not far away. Occasionally he would swoop down within a few yards and I got splendid views of him. The only thing to do was to search over a track at least 300 yards square—almost a hopeless task as the sedge was tremendously thick. But this I did, and after searching for more than three hours it suddenly seemed to me that a great piece of the fen came away just in front of me. Up rose the great brown hen bird, not a yard away, leaving a solitary young owl of a day or two old blinking in the sun. The nest was just a trampled flat span in the middle of a patch of very thick sedge: lots of pellets lying about and the hind quarters and tails of two half-eaten water rats. The eggs must have been laid unusually early—the middle of April at latest.

Most books give the beginning of May as time for eggs to be laid, so the recent cold weather probably killed the other young birds if there were any. About the middle of the afternoon I saw a very brilliant bird flying about some low bushes. I managed to get close up to it and saw that it was a cock golden oriole in grand plumage. It was a bird I never expected to see in this part of the world and I only hope it may escape being shot, though it hardly will, as it is very conspicuous and not at all difficult to approach.

Not a bad record for one afternoon—a new bird and a new nest which I consider one of the best finds I have ever made.

MAY 12. (*Cambridge.*) Went again to see my young owl. When I reached the nest it was nowhere to be seen. The old one kept swooping round in a great state of alarm, but I was not going away, so after searching an hour I found the young bird squatting down at the bottom of some tall dead reeds about 40 yards from the nest looking exactly like its surroundings. I carried it back to the nest and took six photographs with different exposures hoping that one will come out right. It has grown a great deal since last Saturday.

MAY 13. Had a fine walk along the Peddar's Way to Thompson's Water; a good place for duck but too much preserved. On Black Rabbit Warren I found stone-curlews' nests and a nice nest of peewit. This part of the breck is about the best I have been on. For several minutes I watched two hares fighting in a wood. They stood up and hit each other hard with their forepaws so that the fur came out in clouds; then one would chase the other for a little bit and begin again, grunting all the while. They passed within a few yards of me but were so busy that they did not see me.

MAY 23. Went to Chippenham to look for water-rails but found nothing but some young weasels, quite newly born, squeaking at the bottom of thick sedge. On the road I saw a great many nightjars flying about within a few yards of me. They settled on the road, which was quite an inch deep in dust, and they were all remarkably tame. Whether they were dusting themselves or only picking up insects from the road I could not tell.

JUNE 3. (*Lydd.*) Came to Lydd hoping to see something of the Kentish plover—if there were still any at Dungeness. I got close down to the sea to a place where there was a colony of terns—about four pairs of common and six pairs of lesser—and I hunted for the nests for a very long time without success. Then I hid behind a bank and watched two birds go down. I immediately made for the place and found two nests of the common tern with eggs. This is the first time I have found eggs of the common tern. Going along the coast a short way west, I found another tern's nest with two eggs of quite a different type: in each case there was a slight nest of grass placed among stones a good deal smaller than the eggs. After finding this third nest I sat down in a hole in the shingle to blow the eggs. There were a lot of birds about, so when I had finished I got up very suddenly hoping to surprise any birds that might be sitting near. To my astonishment a stone-curlew got up and flew away not 30 yards from where I had been sitting. I ran towards the place, nearly treading on a lark sitting on a nest, and in a few minutes found the two eggs—fresh—lying among the pebbles. There was absolutely no semblance of a nest, not even a depression, though within 20 yards or so there were several scratchings evidently made by the bird.

JUNE 4. To-day I started in the direction of Dungeness and after going over about a quarter of a mile of shingle I came to some big open pools about 200–300 yards long and surrounded with reeds and swamps. One of these pools was inhabited by a large colony of black-headed gulls which got up and made a tremendous din when I came near. They seemed to be nesting amongst the reeds and on one or two small islands, so I took off my clothes and tried to get at them, but the bog was so bad I couldn't reach them. A few birds showed a good deal of excitement when I walked about the shingle itself, so I searched carefully and in about two hours found five nests (3, 2, 1, 1, 1), all with fresh eggs and all new to me. There were small patches of grass and moss scattered about the shingle, and in most cases the nests were placed just at the edge of the grass: small chips of reeds and a good deal of grass made the nest. I then set out for the shore a little west of the lighthouse, where I expected to find the Kentish plover, and soon came across two pairs. There was no difficulty in identifying them—smaller than a ringed plover, lighter on the back, and with the ring incomplete in front. They were in a considerable state of alarm and flew round me crying 'weep, weep', a noise very like a much exaggerated willow-wren. I lay down and watched them, and then searched for the nest but found none. I don't think they could have had young birds anywhere or they would have shown a good deal more alarm than they did. Unfortunately the sun is so hot that the birds are probably right away from their nests, as shore birds so often are on hot days.

JUNE 6. Again to Dungeness and searched on the beach—east and west of the lighthouse—for the Kentish plover. After several hours I came to a spot about half a mile west of the lighthouse and about 200 yards from the shore. There a pair of K.P. flew about me in a tremendous state of excitement. They ran along the shingle in front of me and did all they could to lure me away, but I lay down on the ground and concealed myself behind tufts of grass and low whin-bushes in two or three different places, until after two hours' watching I made out a spot where one of the birds repeatedly went and ran about in a curious way and then came flying about me again. I marked the spot, searched carefully, and succeeded in finding first one, and then about 2 yards away another Kentish plover—just hatched, crouching among the pebbles and most perfectly concealed. They are a good deal smaller, and with their black markings, which harmonize very well with the dark shingle just there, a good deal more difficult to see than the ringed plover.

I often noticed that larks near the sea are inclined to imitate the call note of the ringed plover, introducing it into the middle of their song. Here, while I was watching the Kentish plover, I several times heard a lark imitate the curious 'wit wit' which is the call or alarm note of the Kentish plover.

On a barn on the marsh I saw some starlings sitting, and they imitated most admirably the call note of the ringed plover, and the sort of bubbling note which the ringed plover makes when it flies about chasing another.

Had a very good view of a turnstone in splendid plumage running about the shore with a pair of ringed plover. I suppose he will be off northwards to breed in a short time.

JUNE 28. (*Exmoor.*) Went to Clovelly, pushed my bicycle up the street with great difficulty, then rode to Hartland Quay where I put up at the hotel there. Continued my friendship with Oatway, the landlord, and sat outside talking to him all the evening. I wished I could stay here, but the next morning had to start away early. When I asked for my bill, Oatway refused to give me one and insisted on treating me as his guest; Mrs Oatway came and insisted too and I nearly quarrelled with them, but was forced to give way. He is quite a mystery to me—I never heard of any innkeeper doing that sort of thing before. I bicycled through Bideford and lost my way by trying a new road to Barnstaple, but got there eventually in time for a late lunch. I then rode up the Yeo valley—very lovely, but much spoilt by that arch-villain George Newnes. Up Loxhore hill, about 1100 feet, and on to Challacombe where I drank beer in a quaint little pub, and heard all the Exmoor gossip—about Harry Ridd, 'who weighs eighteen score and is an ungodly sight', and Lord knows how many other famous men! Then rode on to Simonsbath over a spur of Span Head, and down a long slope to the inn which I reached about eight o'clock, fairly tired.

JUNE 30. Started early along the Exford road, through Muddicombe right over the moor to Hawkcombe Head and down to Porlock Weir, and there I lunched with the Grey Ladies.* I sat in the garden till about three o'clock when it began to rain, and then set out hoping to get as far as Bridgwater, only the chain of my bicycle broke soon after leaving Porlock so I had to walk into Minehead in soaking rain and did not get home till midnight.

In August of this year A.F.R. went to climb in Skye with his friend Vivian Le Neve Foster. Writing from Sligachan he says:

We have been doing the Ladies' Chimney and others on the Eagle's nest. The Ladies' Chimney we found blocked at the top with a big boulder. I could just get my head through and no more, so we had to come down again. Afterwards we shifted the boulder from above so that a man could just get through. The next day I had one of the finest walks I have ever had. Up Glen Sligachan over Druim Hain and down to Coruisk where we had a bathe in the loch—ice-cold water and blazing sun. Then up over the slabs to Coir' an Lochain on to Sgurr Coir' an Lochain—down Coireachan Ruadha—over the head of Coruisk to the Bealach na Glaic Moire—down into Tairneilear and across Coire na Creiche. Home over Bealach a' Mhaim—in all 11½ hours. Many golden eagles, merlins and ravens.

Slacked next day, and then went up the western (highest) peak of Mhadaidh from Tairneilear and Coire no Dorus; back along the ridge of Thuilm and over Bealach a' Mhaim.

* These were two very charming old friends of the family.

When we got back we found several new arrivals. Norman Collie, Major Bruce (great climber), and with him Havildar Hurkabir (Hurkir) of the 5th Ghurkas and one of the wiriest little men I ever saw. Collie says he is one of the best rock climbers that ever lived. He was with Manners Smith when he got the V.C. on the frontier, and won the native equivalent himself.

Saw a bird which I am almost sure was an Alpine Accentor. It was creeping about on some big blocks of rock about 250 feet above the sea—not a bush or a tree within miles; very distinct call note.

SEPTEMBER 3–10. Raining hard and I have been learning much about mapping from Norman Collie.

From a camp at Glen Brittle we went to Coire Labain and on to the top of Sgumain, then along the ridge over Alasdair, Thearlaich and Mhic Coinnich; a very fine climb. On to the top of Dearg where we meant to climb the Inaccessible, but it was blowing such a gale and the rocks were getting so dangerous by the rain that we gave it up and came down Corrie Banachdich. This is by far the best ridge climb I have done hitherto.

Just heard that Hurkabir has carried 60 lb. of baggage over Druim Hain and down Harta Corrie and Glen Sligachan with bare feet! He went to the top of Glamaig and back (2460 feet) barefoot in 55 minutes; 37 minutes up and 18 down, coming in perfectly fresh afterwards. In the evening he did all manner of tricks—chopping pennies in two with his khukri—turning himself inside out—and many other things which we tried to do but failed miserably.

SEPTEMBER 11. The Collie party have left and we miss them sorely. Have done a good deal of bug-hunting and found a number of very nice-looking larvae—of I don't know what—on some elders; hope they may turn out to be something good.

SEPTEMBER 13. It's a fine day at last, so John Mackenzie, V. and I went up Sgurr Dearg, reached the Inaccessible Pinnacle in just four hours and proceeded to climb it; John leading—V. in the middle—and I last. The start is fairly easy but higher up it becomes frightfully narrow—a regular knife edge, and very rotten rock. Got up in about half an hour, then down the short steep side—I leading, V. next, and John last. The first pitch is all right but the second puzzled me: the footholds are so very smooth, all basalt dyke and sloping outwards. Good climbing every foot of it and plenty of drop—120 feet on to scree on one side and on the other a drop on to 700–800 feet of steep slabs.

The mountain shadows and blue sky here are such as I have never seen before.... These are days to remember for all time.

OCTOBER. Back in Bethnal Green. The leaves are rapidly disappearing from the trees, but they don't turn a good colour here in London, only make a horrid mess on the pavements, so I shall be glad when they are gone.

NOVEMBER 13. Stayed at Slapton Sands. Ravens fairly numerous all along the coast from Dart to Prawle. A few peregrines and kestrels. I bathed in the sea on November 12, and saw two swallows. No sign at all of bearded tits which used to breed in Slapton Ley. Saw a shore-lark on the strip between Ley and sea; only once have I seen one before—in Lapland in '97.

In January 1900 A.F.R. went to Carbis Bay near St Ives, and had one great walk to Gurnard's Head and back by moonlight. He thought Penzance a mean place—hardly any birds at all.

(*To J. H. Clapham.*) You may laugh at me as much as you like for going to a place where you can bask in the sun in January, but you could not have resisted the delights of Cornwall if you had shared a few of my walks to Gurnard's Head. I will not try to tell you what the sea was like, or the sky, because I simply cannot; but he must have

> The soul of a clod who thanks not God
> That ever his body was born

on a day like that. I stayed looking and looking till sunset and then it was moonrise before I could get myself gone. Ten miles home over desolate granite hills, past ruined mines, stone circles and dolmens. I think I saw more ghosts in that 10 miles than in all my life before—till suddenly coming through a wood lay the Bay of St Ives at my feet, crowded with hundreds and hundreds of fishing boats, each with its light, hurrying backwards and forwards and round and round like in some mad dance. Surely Alfred Tennyson was never at St Ives or he would have said the Pleiades twinkled like the herring boats at night in St Ives Bay—but perhaps he could not have fitted them into the line.

Out of my window at night I could see the Trevose light flashing 40 miles away up the coast, and if I felt more than usually drunk I could see the Hartland and Lundy lights round the corner 40 and 50 miles beyond; drunk I should say not with wine but with the very joy and pleasure of being. I sometimes think I am horribly selfish in enjoying myself so intensely when I get into the country. It is that old Lotus-eater feeling, but for all that I don't think I would part with it for the world.

> So be it when I shall grow old
> Or let me die.

To-morrow, I am afraid, will see the death of that side of me as I go back to Bonner Road. It is satisfactory to have begun the last lap of this M.B. race and to feel that the end is somewhere on the horizon. I wish I could see where it is going to lead to, for do what I will I cannot for the life of me see that brass plate looming in the distance. I have not even dreamt of it, so I begin to think there can never be one. What then?

FEBRUARY. Have been offered the secretaryship (under the Treasury) of a Committee which has been appointed to inquire into the management of the herbaria at South Kensington and Kew. It would certainly lead to further employment by the Treasury but I don't think it would be congenial work and so I refused it. I wonder if I did right, considering the very problematical condition of this bird in the medical bush?

MARCH. Had a letter from F. Gayner who is with Charles Rothschild at Wady Halfa. Lucky dogs. Confound it all, why must I stay bottled up in this cursed London! Many times a week I would gladly rush away anywhere if only I had the means. Lucky perhaps that I have not got 'em, as I am really keen about this medical business and mean to go through with it—London or no London.

MARCH 22. (*To F. Gayner.*) A letter of R. L. Stevenson, which I was reading not long ago, begins somewhat to this effect as far as I can remember: 'I have the strangest repugnance to writing: indeed I often get myself nearly persuaded into the notion that letters don't arrive, in order to salve my conscience for never sending them off'.

I entirely sympathize with him—as I daresay you knew already—having probably received not one letter from me in the four years of our acquaintance. The truth is I don't write letters to anybody—never did—and probably never will. This is merely an acknowledgment of yours from Wady Halfa, most welcome though at the same time most provocative of envy. I read your letter and think of sun, clear sky, roasting heat, bearded Arabs (at least I imagine they have beards), wheatears, martins, shrikes and ducks; then I look out of the window and see gas-lamps, fog, snow, rain and mud, Arabs of the street and smutty sparrows....Your account of the birds interests me a great deal. Most of the species are quite new to me so I shall much look forward to seeing your skins. You will do well to bring home a good series of the various wagtails and wheatears that you come across; they, especially the former, are the very devil to identify....My little doings, supposing I had any to tell you, would sound very paltry and monotonous, and so indeed they are—nothing happens except that I get older and my temper gets conspicuously worse....Three or four years, thank heaven, will see the end of the worst of it. I have quite made up my mind to move from here as soon as I can manage to pay my bill....You might tell N.C.R. that he erred greatly when he said that I could probably make myself very unpleasant if I wished. I regret to say I have occasionally tried to do so (when people have irked me very badly) and have failed miserably. I only wish I could....

P.S. By the way, unless you have done so already, read R.L.S.'s letters. As a rule such things bore me intensely, but I have revelled in these two volumes—which are big, but not half big enough. Such in every way delightful letters were never before written by anyone. The curious thing is that as you read them you get to know the man so intimately that you could swear that they were written to you, and not to some happy person that you have probably never seen. There is evidently no effort made in the writing of them, and in many the grammar is vile and the construction worse, but they are so spontaneous and so witty that they are like to be a joy for ever; read them.

In April A.F.R. went to climb in Cumberland. At Carnforth he joined a party of others 'and we had a grand drive by moonlight to Wastdale Head where I had never been before, and truly the half had not been told me....Climbed two gullies on Great End and had my first experience of the use of the ice axe; a jolly glissade from the top of Cust's Gully'.

APRIL 12–15. Climbed Scafell Pinnacle. It took more than half an hour cutting steps to get across the head of Steep Ghyll—usually a simple walk of about 10 yards; then down the Professor's Chimney, cutting steps all the way. Deep Ghyll was one smooth slope of snow from top to bottom; there was at least 30 feet of snow in it and all the pitches were buried.

Climbed the East and West 'Jordans'—the most difficult bits of rock climbing I have done, and went up the Needle arête. Did not attempt the Needle itself owing to terrific wind. The first piece—a slab sticking from the base of the Needle—is decidedly difficult. This was a splendid holiday and a great increase in my knowledge of the mountaineering craft.

APRIL 21. Made my first acquaintance with Essex. I went to Beadwell Quay, a quaint little place at the mouth of the Blackwater in country very much like the fens but no sedge. Glorious day, blazing sun. Sat on many gates and enjoyed myself hugely. Gossiped with many natives—nice people all. I asked one man to whom the land hereabouts belonged. 'I am told it belongs to Canon Gregory of Paul's Cathedral Company; a lot of high-minded parsons I believe they are but I don't rightly know.' Walked round the sea wall to near Southminster and saw good salt marshes which should be full of birds in the winter. A pair of birds got up out of a pond and I am pretty sure they were greenshanks.

JUNE 2. Went to Tring and saw one of the prettiest nests I have ever seen. It was a pochard's under a gorse bush. The old bird flapped off the nest almost at my feet and left the eggs exposed. There was a good deal of down, but not as much as there will be by the time the eggs are ready to hatch. I saw one male shoveller but the nest has not been found yet. They generally build their nests a very long way from water —sometimes 2 or 3 miles.

I also saw the nest of a tufted duck. It was placed on the side of a bank under a thick growth of nettles and the eggs completely covered with dead leaves.

JUNE 9. Went up to Scourie, Sutherland. Heard a pair of greenshanks making a low whistle like a redshank: a sort of 'tak' rather like a snipe—a very curious rippling whistle it was. Found their young birds just hatched among heather and rocks about 100 yards from water. I believe the cock stays right away while the hen is sitting, and she sits as tight as a stone, especially near the time of hatching. As soon as the young are hatched they make no attempt at concealment. I watched a pair of golden plover with young. They have a wonderful variety of alarm notes and they played all sorts of tricks to entice me away. This is the first time I have ever seen golden plover in their breeding places; they are rare about here and I have only seen one pair since I came. Morrison took me out one day on to the far hills beyond the Chain Lochs where the divers breed every year. Found their runs and 'forms' but they had not eggs yet—too early for them so high up as that.

I left Scourie and went to Ravenglass, Cumberland. Took a boat over the mouth of the Esk to Drigg Point hoping to find Sandwich terns. It is a very awkward place to walk about in and the gullery is strictly private. The birds make a din which can be heard for miles. By great good luck I walked into a small colony of Sandwich terns and found four nests with fresh eggs. The bird has a very peculiar short yapping cry and a way of flying very straight towards you just a few yards above your head, which is a thing I never saw the common tern do.

JUNE 15. Went to Buttermere and then to Grassmoor to see if the dotterel is still there. Heavy rain and dense clouds, and I was just thinking of giving it up when I saw a dotterel running along the ground in front of me. It looked as if it might have run from its nest, so I searched for more than two hours, getting very wet and cold, but found nothing. Anyhow it is satisfactory to know that the bird is still in the district.

JUNE 20. Arrived Holyhead at 3.30 a.m. hoping to go to-day to the Skerries to find the roseate tern. It was very foggy all the morning so I had to content myself with a walk along the shore to the South Stack. Did not see very much on the way there, but when I got down to the edge of the cliff I found myself within a few yards of a great breeding place of guillemots, razorbills and puffins—thousands of them literally shoving each other off the ledges trying to find room to sit. The razorbills seemed to be in the deeper holes in the cliff rather than on the ledges, while the puffins were on the steep banks above the sheer cliffs, and busied themselves running in and out of their burrows. I came across a colony of herring gulls and their nests were quite accessible —either on a patch of loose scree or on grassy ledges.

Bicycled 18 miles to Cemmaes, on the north of the island, where a boat was waiting. Sailed in about one hour the 8 miles to the Skerries. Immense flocks of terns were flying everywhere and it seemed hopeless even to try and identify the roseate terns in that one afternoon—much more so to try and find their nests. Fortunately one of the lighthouse keepers is a good naturalist and knows the birds well. He soon pointed them out to me, and as far as I could judge there must be about fifteen pairs of the R.T. on the island. Their flight is more graceful and more powerful than either the Arctic or common tern. The wings are appreciably shorter and the tail longer. I found it hard to see either the rosy colour on the breast or the black bill and I wondered if I should have the luck of finding their eggs. It was difficult enough to watch one of the birds down to a nest, but when it came to knowing for certain which of the many nests the bird got up from, I was fairly beaten: the ground was literally covered with eggs. However, after repeatedly watching birds get up from the same place several times over, I made certain of two nests—both with eggs. This is probably the rarest nest I have ever found. It was fun to see the roseate terns chasing the others and making them give up their food in the same way that skuas do.

The next day I wasted some time between Holyhead and South Stack looking for corn buntings' nests. I have never managed to find them; the great difficulty being that one does not like to wander about in the mowing grass at this time of year. On the south-west side of Holyhead mountain I saw a sight which pleased me immensely. I was climbing down a small cliff when a pair of peregrines came and flew backwards and forwards in front of me—shrieking and making a tremendous fuss. I sat very still, thinking the nest must be near me, but they went away, and in a short time came back with three young birds nearly fully grown and well able to fly; all five flew about me and settled quite close on different points, giving me better views than I ever had of peregrines before. It is satisfactory to know that they can manage to bring up young birds within a mile of a town of 10,000 inhabitants.

Part of this month I spent in Ireland, and near Belmullet, in a marshy place running down to a small estuary, I saw some red-necked Phalaropes and found a great number of nests. They are mostly placed on dry ground where the water does not stand—in grass just sufficiently long to make the eggs difficult to see. Dunlin are in rather wetter ground. I walked over Annagh marsh, and there the Phalaropes were as tame as can be, swimming about the pools within a yard or two of one. At Achill, on the west side of the island, are piles of big stones set out on a large shingle bank, and you hear as you walk about a curious sound between a croak and a rattle. Bending down you smell a peculiar smell, and when with great labour you lift up the great stones, you find crouching amongst the lowest stones two stormy petrels. I unearthed, or rather unstoned, several pairs but none had an egg.

IV

EARLY TRAVEL

Diary kept by A.F.R. when he went to the Dolomites with Vivian Le Neve Foster.

AUGUST 10, 1900. At Munich we strolled about the town and I bought myself a hat of the kind affected by German tourists in the Alps, but without the cock's feather; not unlike a pudding basin—dark green and distinctly bounding, but it was pronounced to be 'sehr hübsch'. We arrived at Zell am See in pelting rain and ran to the first hotel we could see. It was the 'Post'; an old-fashioned sort of place with stables on the ground floor and a nice horsy smell in our bedroom at the top; outside there was a good smell of wet earth, wet pine trees and cows, and I knew that we had got to the real beginning of our journey.

AUGUST 12. We started on our southward walk through swampy ground to the village of Bruck where we crossed the Salzach river and the broad Pinzgau valley. Then up the narrow Fuscher Thal to Fusch, where the road narrows and climbs a gorge with a tremendous torrent roaring at the bottom; later the valley widens into green pastures. On the east side are some marvellously shaped bright red rock peaks; on the south and west are the huge walls of the Tauern Range—all covered with snow and dazzling red in the setting sun. We put up at the Tauern Inn, and spent the evening in the native way, with music, dancing and drinking. It was a Sunday evening and everyone came in in their finest clothes and I think their thickest boots. The shape of the room was something like the letter 'T', encumbered with chairs, tables and barrels, and not much more than 20 feet long; for all that, it held some thirty or forty people with generally a dozen or so dancing at the same time. Some of the dances were very graceful and many were exceedingly complicated. The musical instruments consisted of a French horn, an accordion and a piccolo, and the repertoire of the players was almost unlimited. They played their different parts without music and without hesitation, and whenever one of the players got tired or thirsty he handed his instrument over to the nearest man standing by, who took it up quite as a matter of course. A young man and a girl sang a duet which caused a great deal of amusement, but unluckily we could not make out the meaning of it. All this—in a little room lit by three or four candles, heavy with tobacco smoke and the steam of cookery—made a scene which I shall always remember.

Up at 4 a.m. and off with a guide who was dancing till one o'clock this morning. He didn't seem to be much the worse for it to judge by the terrific pace with which he started away. An hour's hard walking up the valley brought us to the Gasthof zur Trauner Alpe, where we laid in a stock of provisions for the walk. A very well-marked path led us to the foot of the glacier, and from there it would be impossible to miss the way (in good weather) over the low pass to the Glocknerhaus. Looking north from the lower part of the glacier we saw the rugged chain of peaks right over Zell, and all the nearer Tauern peaks, to the Bärenkopf and the Spielmann. An easy snow grind took us over the Pfandelscharte pass (8745 feet), and starting down on the other side we soon came in sight of the splendid Gross Glockner standing up like a pyramid above all the other peaks, with the great black cliffs of the Glocknerwand in shadow to the north of him. Half an hour's running down over snow and rock brought us to the Glocknerhaus, a large hut belonging to the German and Austrian Alpine Club, with a splendid view of the snout of the Pasterge Glacier and the Gross Glockner. We lay in the sun on a bank of gentians looking at our mountains till about two o'clock, then we started off with two guides (I am sure we could easily have dispensed with one) for the ascent of the Gross Glockner by the Hoffmann's Weg. Up the east side of the glacier for about 2 miles, then across—about a mile and a quarter broad—with very few crevasses and none too wide to jump. Along a steep snow slope on to loose rock on to the Glockner Kar glacier, and then to the Adlers Ruhe Hütte (11,370 feet). Southwards we could see perfectly clearly every peak and rock of the Dolomites; westwards—where the sun was setting—we saw the Ortler, Adamello and the Oetzthal. I could have stayed there all night but for the cold which was intense and a bitter wind which drove us into the hut. It was a three-roomed hut, and it held twenty tourists, twenty guides and a few odd natives, all cooking and drying clothes and otherwise adding to the frowst. I have been in some stifling dens in Lapland, yet I don't remember any atmosphere that could compare with this. Most of us smoked and sweated and steamed near the fire, while the lucky few drank warm wine and nauseous pea soup; but it was huge fun and we enjoyed it vastly. About eight o'clock we retired to bed in a room some 18 feet long with ten narrow coffins on each side and a narrow gangway down the middle. The twenty guides climbed up into a sort of cupboard at the back of the kitchen chimney and consoled themselves with drink and tobacco; at all events they must have kept warm, which was more than we did. There were some blankets but they were wringing wet and no use. I borrowed as many pairs of stockings and shoes and cast-off raiment as I could, so slept tolerably.

AUGUST 14. Up just in time to see the sun rise—such a sunrise, I cannot describe it. Along steep snow to the base of the rocks, where we put on the rope and had half an hour's really good climbing between the Klein Glockner and the Gross Glockner. There are some iron ropes in places but these rather add to the difficulty than otherwise, and should be avoided. A quarter of an hour at the top (12,460 feet) was all too cold, so we turned our backs and came down at a great pace, racing over the snow slope and glacier to the Glocknerhaus; then a dawdle back to Heiligenblut.... A merry evening drinking to the health of the Gross Glockner.

AUGUST 15. Awakened by a great ringing of bells and firing of guns, which went echoing up and down the valley, making a tremendous din. It was to celebrate the Feast of the Assumption, and by nine o'clock the whole village, dressed in their sombrest garments and singing and praying, processed from the church to a shrine about a mile down the valley and back again. I thought what a large number of the people looked half-witted or even worse. I wonder if it is the same in all small valley villages. The last time I saw the keeping of the Feast of the Assumption was at Brissago in '93, and there the people wore their finest clothes, the brightest handkerchiefs over their heads, and many were exceedingly handsome; a great contrast to these ugly people with their black straw hats and long black trailing ribbons. In the church here are preserved a few drops of the Holy Blood in a bottle enclosed in a vessel 42 feet high, and there is a pretty legend concerning it which is told I think in Miss Amelia Edwards' *Untrodden Peaks*.

Strolled about in the groves of alder by the river and along the maize fields. Stopped on a road to play with some boys who were rolling wooden balls. I don't know what their game was, but they had no notion of using their hands. When I threw a ball for one boy about 5 yards away, he ducked and was nearly hit on the head. They thought we were very clever to be able to throw a ball and catch it, and it is the first and will probably be the last time that I have ever been applauded for my performance with a ball.

In the evening we dined off delicious trout fresh from the river, and later sat with the musical society of Winklern in the common room of the inn. About a dozen men and women of various ages, and two priests, sat round a big table drinking wine and singing duets, trios, quartets and choruses—all in perfect time and tune and generally from memory. Now and again one of them would sing a sort of jodelling solo while the rest sang a humming chorus. It was a concert that one would go a good way to hear in England, and yet they were just a haphazard collection of peasants who chanced to be in the inn together and sang to amuse themselves and pass the time.

Here, as at all the other inns we have stayed at in this country, we have made out our own bill on leaving; a pretty certain test of the remoteness of the country.

AUGUST 16. Up the hill to cross the low ridge which divides the Möllthal from the Pusterthal. Near the top—at a little inn where we took shelter—we found ourselves on the edge of the Italian-speaking Tyrolese, the padrone talking the most extraordinary mixture of German and Italian. About an hour's walk brought us out into the broad fertile Pusterthal; then on through hay fields and maize fields over the Isel into Lienz. Lienz is a sleepy little town and the people much more Italian than German in appearance. As it was getting rather late and as our road merely followed the railway in all its windings we decided to train to Innichen.

AUGUST 17. Started along a road between fields purple with colchicums to Toblach; then turning south up the Höhlensteinthal we found ourselves in the Dolomites. Up the valley to Landro, and there towering above a little valley on the left were the three huge fingers of the Drei Zinnen.... A storm was coming up so we decided to stay at Schluderbach, and very glad I was that we did, as here I witnessed one of the most magnificent storms I have ever seen; such a wonderful clearing up afterwards—a great rainbow stretching from Piano to Cristallino and completely framing the Cadini in the light of the setting sun.

AUGUST 18. Strolled up the Val Popena into Italian territory to-day and rounding the north-east spur of Monte Cristallino came upon the Lago di Misurina. To our disgust there was a huge hotel built at the lower end of the lake. But for this vile abomination I think Misurina would be one of the most beautiful places imaginable. We took rooms at an old albergo, and while Vivian went to the top of some fir woods I was more lazy and searched out a snug resting place in a bed of whortleberries and heather, and there spent four blissful hours. I dozed and read and looked at the mountains. My book was *Tess of the d'Urbervilles*, but not even Thomas Hardy's best descriptions of the West Country could make me wish to be anywhere than where I was.

AUGUST 19. Took a boat to bathe from on the other side of the lake; after this we filled our pockets with bread and chocolate and walked up the valley to the ridge that joins the Drei Zinnen and the Cadini. We skirted round the south side of the Drei Zinnen over huge screes and slopes white with edelweiss, till turning a corner we found ourselves at the foot of the Kleine Zinne. It is a stupendous rock—a spire 800 feet high—looking from this south-east side quite perpendicular if not actually overhanging; as if a moderately strong gust of wind would blow it into the valley below. Whilst we were sitting 'over against' it and marvelling at the Kleine Zinne, a party of German tourists came by, never raising their eyes, but anxiously plodding along to their lunch at Misurina. I don't believe they saw a yard beyond their feet—they most certainly did not see the Kleine Zinne.

We continued up a steep slope covered with yellow poppies and purple gentians until we came suddenly in sight of the north face of the mountain. This side of the middle and west peak is if anything even more terrific than the Kleine Zinne

—immensely bigger, and with a great snow-filled rent separating the one peak from the other. At the foot of the precipice are several hundred feet of screes too good to be missed, so we raced down and found ourselves in the rubbish-heap of the mountain—a wilderness of blocks of Dolomites varying in size from that of a piano to that of a good-sized house—all tumbled together in the wildest confusion. Some of these boulders were so temptingly placed that we sent them rolling down the hillside hoping that nobody was below. We arrived back at Misurina having completely encompassed the Drei Zinnen. It is not a beautiful mountain, but it is more grim and awe inspiring (awful in the true sense of the word) than any mountain I have seen.

AUGUST 20. Very sorry to leave Misurina, but there are too many people about to make it a place to stay at for long. At the hotel at Tre Croci we had great difficulty in getting rid of a Mrs E. and 'my dear girls'; she wanted us to walk to Cortina with 'my dear girls', and should we find ourselves in any difficulty 'be sure to call on my husband, the Reverend E., the mountaineer, you know'. We luckily escaped before I had said anything very rude and walked down to Cortina alone and unattended.

AUGUST 21. Started along the Falzarego, under the grand red buttress of Tofana. Up through pine woods, occasionally catching glimpses of the Croda da Lago standing up like a huge bare black cockscomb, and through woods full of nutcrackers sitting on the tops of the pines chattering and laughing at us as we sweltered along under our rucksacks. Coming out above the trees into the open we found quite a little colony of people encamped for the haymaking. It is difficult to see what the hay consists of except boulders and an occasional gentian; but these people come up several thousand feet, so it must be worth their while. I unintentionally annoyed them by walking on a patch comparatively free from boulders, (it must have been a hay field though I could see no grass) for they shouted and would not be pacified until I had left it.

We made a slight detour to the right to see the Cinque Torri—five most fantastic rocks perched up like huge monoliths on the Averau plateau; the southernmost tower by far the biggest (about 600 feet high I believe), and apparently quite inaccessible. There is a curious chimney running right up the middle of it which can be climbed, and we meant to climb it, but it was so hot and we were so lazy that we contented ourselves with inventing ways of climbing the towers. Some I believe have never yet been climbed. There were a great many plants growing among the rocks which I could not make out, and like a fool I had not brought any book with me. Up over great slabs of limestone to the Saddle (7875 feet) between the Nuvolau and Monte Averau, and there we sat and devoured our lunch, with a view of entirely new country before us and a view fit for the gods. There were still a great many miles between us and Caprile, so we dawdled down the steep hillside, and with many halts for as many pipes eventually found ourselves in the Val Fiorentina and Caprile. Caprile is a squalid little village but it has an interesting and well-wrought Lion of St Mark on a pedestal, which shows that we are again in Italy. I like the copper water-buckets used by the people, beautifully hammered and looking as if they had been in use for centuries.

AUGUST 22. A thunderstorm last night, and to-day the river is twice the size it was yesterday and full of a bright brick-red mud. Numbers of men were out with huge hand-nets but they didn't seem to be catching anything. A few miles below Caprile the river runs into the beautiful little green Lago d'Alleghe and here we hired a 'barca' of great age and greater weight, which we propelled gondola-fashion with two tiny little paddles, and made our way to the west side of the lake where we had a delicious bathe—blazing sun and ice-cold water. Afterwards we sat outside a snug little inn and devoured risotto and other things. I pleased the hostess a good deal by changing for her a perfectly black shilling which she had received in payment from an Englishman many years ago. I should be sorry to say how long we took to cover the 5 or 6 miles down the valley to Cencénighe (it was very hot and there were a great many tempting wayside resting places), but we arrived there eventually. I wonder why so many of the places hereabouts break the usual rule of pronunciation, or rather accentuation, and call themselves Ágordo, Cencénighe, and Álleghe? After leaving Cencénighe we went to Forno di Canale, and put up at a little inn—rather grimy—where they gave us a very good dinner, charged us very little, and our beds (so far as we found out) were untenanted.

AUGUST 23. Drizzling and the clouds very low when we left Forno and went along the road as far as a village, where we got rather mixed in finding our proper path but were eventually put right by an aged hag who assured us 'that it was a very sad life and that her husband had recently become a corpse and had gone to the church'. Our way was a stony track, more like the bed of a not quite dry stream than anything else. Occasionally we met ox wagons bumping down laden with hay. They possess two wheels in front only, and sticking out behind are two long poles which act as brakes and steering gear. The driver walks in front, or hangs on to a long pole which sticks out between the heads of the oxen; the strain on their necks must be tremendous. Higher up our track turned into a water-course, and we came out on to the top of the Vallès Pass, and once more into Austria.... At San Martino di Castrozza—the place I have been aiming at ever since we started—we took a room at the most modest of four hotels and then went out for a stroll. At dinner in the evening we talked to a German and first heard of the taking of Pekin which happened ten days ago! So we have been a good deal out of the world lately.

AUGUST 24. The sun out for an hour or two and we saw what a beautiful place is San Martino di Castrozza. I should dearly love to stay here and do some climbing, but it would be very expensive as the guides' fees are heavy, and even our little hotel, which is the cheapest in the place, is much more expensive than we expected. In fact I have quickly come to the conclusion that this is no place for the poor tramp, so we must pack up our sacks and start off for Primiero.

It has begun to drizzle, now it rains, and now it pours in torrents. Dozens of salamanders come out of their hiding places, and in a few minutes our track is covered with crawling yellow and black beasts. We arrive drenched to the skin at the 'Aquila Nera' at Primiero—one of the true old-fashioned inns (without the fleas), kept by one Bonetti, a most courtly old gentleman who waits on us himself at an excellent dinner of trout and omelette. He shows us with great pride a copy of Miss Amelia Edwards' *Untrodden Peaks*, in which mention is made of his inn; he also has a copy of Gilbert and Churchill's *Dolomites*.

At every half-hour during the night a watchman goes his round and shouts in a very aged raucous voice, 'Brutta notte; niente di nuovo'.

AUGUST 25. (*Trent.*) There is an air of stateliness and departed ecclesiastical grandeur about this town which pleases me a great deal, but it is sad to see the fine old mansions and palaces used as warehouses, and all falling to wreck and ruin.

(*Torbole. To his Father.*) I have just had a swim—such a swim—as good as ours at Brissa or even better. I had learnt from you that this was a beautiful place, but the half had not been told me. We got here about two hours ago and my heart has quite gone out to it. All the mountains are perfectly clear, and what a colour! I can see the broad flat plain at the end of the lake and the dark line of the breeze coming up. There is only one thing wanting here, and that is written on the other side of this postcard, and Mother. Here comes the wind—all the reflections are gone but there are new and wonderful colours....Do you know that I never saw olives until I came here? How jolly they look when the wind catches them and turns up the leaves.

AUGUST 27–SEPTEMBER 2. How the week went by I cannot tell. I have a confused recollection of many glorious bathes in the lake, visions of peaches and green figs, and wanderings in vineyards and oliveyards. It was a week of lotus eating and contentment; perfect sunny days and moonlight nights. In the morning a bathe and a bask in the sun; in the afternoon a walk over to Riva and another bathe in the shadow of the hills, with afterwards a cup of coffee under the colonnade of the market place.

Torbole is an ideal place for a lazy holiday; you can walk all day or sit all day and you never get tired of either. There must be something in the atmosphere here that makes people cheerful; the fishermen play bowls and eat melons all day long, and the women washing by the lakeside laugh from morning till night; it is certainly a place to come to again.

SEPTEMBER 3–5. Went by a queer little steam tram to Brescia. The 14 miles took nearly three hours, but it was a pleasant journey and many of my fellow-passengers were amusing. At Brescia I had some difficulty in finding the object of my visit—the museum where is a certain bronze statue of a winged Victory. I saw a photograph of it once at Chapel Knap, and made up my mind some day to see the original. I found it at last and it is enough to say that I was not disappointed; I would go to Brescia and back to see it and nothing else. The custodian of the museum got tired of me about midday (or else he was hungry), and cast me out, so I wandered into the town and found the market place where they were selling green figs at twenty-five for a penny. The next half-hour can be better imagined than described. On going back to the station I saw a train labelled 'Lecco'. It was very alluring and not to be resisted. When I arrived at Lecco the town was indulging in a splendid funeral. It was an impressive sight, but the three brass bands in different parts of the procession all played different tunes at the same time, which rather marred the effect.

(*Milan.*) Sought a bed at a mean and not too scrupulously clean house, for the wherewithal was running low. I spent the afternoon on the roof of the cathedral and by the time I had finished with the view from up there it was too dark to see the inside of the cathedral, so that must wait with many other things till next time.

In January 1901 A.F.R. went with N. C. Rothschild to the Sudan to make a collection of birds and small mammals. Apart from a few rough notes, I have no record of the time spent out there, but an account of their collection of birds made from Shendi was published in the *Ibis* of January 1902. These letters will, however, give some idea of what they were doing.

FEBRUARY 7. (*To F. Gayner.*) My journey out was a great success and I made a few pleasant acquaintances—as usual all of the male sex. All day long it grew warmer and southerner and I forgot Whitechapel and all its works and became a new man. At Port Said it was simply magnificent; what with the flamingoes and pelicans and kingfishers—not to mention the sun and the desert—I nearly went off my head with delight. Everything is new and wonderful and I wander about the streets and stare like a babe open-mouthed at Arabs and Turks and Copts and Nubians and mosques and kites and camels and bazaars and donkeys—but you know all about it so you can imagine my state of mind....

FEBRUARY 22. (*Camp near Shendi. To his Father.*) A day or two ago we made a camel expedition several miles into the desert to shoot gazelle; we saw a good many but failed to get near enough to shoot them. Along the river banks there are occasional groves of palm trees, but mostly it is thick tangles of all the prickliest trees in the world. One of the most conspicuous plants is a thing called the Sodom apple; it has a large fruit like a pomegranate full of white juice which the natives use as a cautery. There is also a very pretty little green and black gourd with a yellow flower which climbs all over the other trees.

We spend all our time collecting birds and bugs and beasts. The fauna is entirely tropical: sunbirds and hornbills and bee-eaters and all sorts of other jolly things. We have already got some very good birds, and possibly one or two that are new. The butterflies are not so very much brighter than the English ones and not a bit more numerous. There are a good many scorpions about and I caught one in the tent the other night as big as a good-sized crayfish, but nobody has been stung by one yet. Snakes are very rare—I have only seen one small one. Crocodiles are hardly ever seen here though they are found in the river both above and below Shendi. The natives bring us all sorts of beasts which they catch—gazelles, foxes, cats, rats, hares, hyenas, and once a huge water tortoise about 3½ feet long....

I am tremendously fit, more so than ever before; almost as black as an Arab and as tough as possible. The Sudan climate at this time of year is as good as we could wish for. We live and eat and sleep always out of doors, and except from about twelve to three—when it is too hot to move—never feel slack. So far, our highest temperature has been 110° in the shade, but that is nothing like so overpowering as 90° in England—probably because it is so dry....

MARCH 3. (*To A. I. Simey.*) The great difficulty here is that there are such a tremendous number of things worth taking that it is hard to know where to begin. You can imagine it is a very strange feeling to find yourself plumped down in the midst of an entirely tropical fauna—sunbirds and weaver birds and hornbills and beautiful long-tailed rollers, and all manner of bright-coloured little finches. It is most bewildering at first. I have found one or two nests, but it is difficult looking for them when one has only just begun to make the acquaintance of the birds themselves.

Lepidoptera are on the whole rather scarce, and those that there are, are not a bit more brilliant than our own in England. Hymenoptera of every kind swarm—ants, bees and wasps in profusion; most of them are probably new and undescribed but we have neither the time nor the knowledge to collect them.

Of crocodiles, there are I believe a few in this reach of the Nile, but though I bathe daily I have not been attacked. I have seen one or two big monitors 3 or 4 feet long which live in the river, but they are much more afraid of me than I of them. We have at present six gazelles alive, four young and two old; they are the most delightful little beasts imaginable and they like to sit on your knee and be nursed all day long. I wish we could bring them home but they would probably die on the way. We have also a young hyena and two baby foxes about the size of kittens, very soft and playful; several bags full of live hedgehogs and a large monitor. Altogether, it is as you may guess a very jolly life, better than one will ever have again. I am afraid it is not a very good preparation for more years of London but that can't be helped....A great flight of cranes is just going over northwards—10 p.m. —making the devil of a noise. Spurwing plovers whistling all round and Egyptian geese konking....

(*To F. Gayner.*) Of birds, our nightjars are perhaps the best. We have five *eximius* including a female never yet described; a male and female of a very beautiful long-tailed species, and a female of a third (to us unknown and possibly new) species, very like *eximius* but greyer. We have a very pretty little-eared owl—an inhabitant of ruined houses, and a splendid big Bubo from a hill 10 miles away. I shot the male bird and found the nest with two young birds in down in a hole in the rocks. I go to-morrow to photograph them and hope to shoot the female bird. A beautiful chestnut-capped swallow breeds locally in the mud banks, and a swift among the leaves of the palm trees—both probably rare birds. Pigeons are very abundant—we have I think six species—and some of them certainly good. One woodpecker and a most brilliant barbet which imitates the form and colouring of a woodpecker—both rare. Two species of fantail warbler—a desert and a cultivated species; two nests of the latter but no eggs so far. Other warblers are very puzzling: there are of course many migrants—some of our own English species, and some others, so it is rather difficult to pick out the resident birds. A blue and red racket-tailed roller is perhaps the handsomest bird we have got; they are scarce here but very likely common farther south. *N. Metallica* swarms: I think we have its nest but am not quite certain. Our larks, wheat-ears, small finches and weaver birds may be good or not—it is impossible to say yet. I am sure we have got a very good collection of the birds of the district even if none of them are new, and after all that doesn't matter in the least. Of mammals we have a very good lot on the whole, considering the difficulties of trapping. Altogether this has been a very great success and it has certainly been far and away the very best time I ever had in my life—I never expect to have another half so jolly. We have the most perfect camping ground in the Sudan, the best tent that can be made, and we are made in every way as comfortable as one could wish for. I think that accounts in a very large measure for our fitness and our good collections.... I wish I could find some berth out here. I should never pine for London or any of its ways. It will be vile coming back....

Back in London A.F.R. notes in his diary that he has had an offer from H. J. Elwes to accompany him on an expedition either to Peru or to Java, but as this would mean 'chucking up the M.B. till December 1902 I can hardly accept; but it is very very tempting'. In 1903 he qualified as M.R.C.S., L.R.C.P., but soon afterwards writes to his father to say that 'medical practice as a means of livelihood does not attract me—in fact I dislike it all extremely'. This dislike comes out many years later in a letter to a friend, in which he says:

...I am painfully conscious of having done a very foolish thing last night when I advised M.T. to give up his idea of going in for medicine before it was too late. I don't know whether he thought I was serious about it—perhaps not—in that case no harm was done, but sometimes a remark of that sort sticks unexpectedly and works as an active poison. I should be exceedingly sorry if that happened in this case. I made such a horrible mistake myself when I went in for a profession which I loathe, and I am so constantly regretting it, that I cannot resist the temptation sometimes of saying nasty things about it. To do so argues me a fool, no doubt, but I am content to risk that so long as I don't hurt the feelings of any other person. If M.T. ever gets on to this subject, you would ease my conscience and do me a great favour if you pointed out that I am altogether a mistake—a very angular peg looking for a suitable hole—and not altogether responsible for my words and actions....

In January 1904 A.F.R. went again to the Sudan with N. C. Rothschild, and this time F. R. Henley was of the party. Afterwards A.F.R. gladly accepted Rothschild's offer to travel to Japan and the Malay States in order to make a collection of moths and butterflies.

JANUARY 1904. Henley seems a very good sort with a pretty gift of humour. He is a very welcome addition to the party—not at all what I expected of a Harrow Balliol athlete....

At Cairo I went and interviewed Sir Horace Pinching,

chief of the Sanitary Department here, with a view to a possible job for myself, but the prospect is rather remote.

(*Wady Halfa.*) We rose early and rode out about 16 miles into the desert to-day to shoot gazelles, but saw none. Tracked a big lizard for about 2 miles through the sand and eventually ran it to earth in a hole at the foot of a small tree; immense Arabic excitement when we dug him out.... Slept out under the stars.

Left Shireik on thirty-three camels and rode to the Wady abu Sellam where we expected to find wild donkeys but there was no sign of life; it was a most God-forsaken spot. ...On to Nikheila, and while waiting for our baggage camel to turn up we had to take shelter from a strong wind by a palm tree. Made a glorious bonfire of the fallen palm leaves, the warmth of which was very comforting. Our camp was on the edge of the Atbara battlefield of April 8, 1898. Beyond the Dervish camp is the river, with a beautiful fringe of palm trees and palm scrub and yellow flowering acacias full of birds....I rode about four hours into the desert with two natives and slept out at a spot where I hoped to shoot wild asses in the morning, but no sign of them. Later came across three but couldn't get a decent shot. Went back to camp by moonlight.

FEBRUARY 3. Hunted for wild asses but could not get near them. At last Charles gave his rifle to a native who was with us and he stalked one very skilfully, killing it at about 100 yards; it was a very old male of great size. Henley and I went out in the evening to see the beast. It was a fine moonlight night and it amused me to think we were walking 4 miles just to see a dead donkey by moonlight. The shikari who had shot the animal was in tremendous high spirits at the prospect of good backsheesh, and broke into a wild chanting of Sudanese songs.

Crossing the battlefield with this shikari to-day, he pointed out some of the wheel tracks of the guns of six years ago. He gave us what must have been a very dramatic description of the shelling of the trenches below, but all we could make out were the two English words, 'battery' and 'Hunter-pasha'. This shikari was taken prisoner at Omdurman but does not seem to have had much sympathy with the Dervishes.

Pottered about collecting birds. The Sudan is beginning to 'stoke up' a bit, and even the nights are hot now.

FEBRUARY 7. Crossed the river on camels and went due west for a few hours and then saw a small troop of donkeys. In the middle of stalking them they were frightened off by some gazelles, but we saw them again later on some ground as bare as the palm of your hand. We again stalked but failed to get a shot. By that time it was sunset so we sought out a bush and dined sumptuously by the light of the stars off sardines, bread, and dates. Had a discourse about stars with the two natives, and greatly astonished them by being able to point out the way to Nikheila, Atbara, Kassala and other places.

A native shikari has just shot two wild donkeys, so now our mission to Atbara is finished, as we only had leave from the Sirdar to kill two or three donkeys. Our skinner went out in the afternoon with an acetylene lamp, skinned all night, and brought back the skins in the morning.

Spent the morning in a tree overhanging the river where Henley was trying to catch little fish on shark hooks. He didn't succeed in catching any, but a big black and white king-fisher (*Ceryle rudis*) sitting on one of the outer branches of the same tree was more lucky; we saw him catch a fish that seemed quite as big as his own head. He sat and chopped and chawed at it, holding it all the while in his bill but every now and then throwing it into the air and catching it again very cleverly. He made three or four attempts at swallowing it, but had to continue with his softening process before he finally got the fish down, and even then the tail stuck out from the corners of his mouth like a beard, until he could get it down with a bigger gulp than before.

FEBRUARY 17. Arrived at Khartum and found a new civilization and a grand hotel where was a rubbish heap the last time I was here. The change since three years is quite amazing. Then there was only one house with a roof—the Palace: now there are a War Office, a huge Post Office, Gordon College, Public Works Office, and innumerable other comfortable-looking houses. There is a well-made road along the bank of the Blue Nile and other roads are being planned out further back through the wilderness of deserted mud houses....Some of the contrasts of ancient and modern are very curious. For instance in the cloth bazaar I saw them weaving cloth with the most primitive loom imaginable, while in the very next booth sat a man working a Singer sewing machine. I went to the Gordon College and saw a venerable man teaching the principles of mechanics to the youth of Khartum, while in the garden at the back of the building was a pair of oxen drawing a plough that might have been used in the time of Moses....

FEBRUARY 28. (*Kerma.*) Rode up to a 'mountain' about 8 or 9 miles north, called Gebel Abu Fatmeh. Within a mile or so of this hill the country looks quite cultivable, but as you approach its bottom the ground is scattered with granite rocks split up and rounded off by the heat and the sand. Some of the rock looks very like fine quartz. The soil here must be richer than that in most parts of the Sudan, melons, pumpkins and onions grow to an immense size.

(*Merowi.*) Met Jackson Pasha, and found him to be a most interesting and genial man who has seen most things in recent Egyptian history. He was at one time commandant at Assuan, and again in command of the Egyptian troops at Fashoda when Marchand was there. He gave an amusing account of K.'s interview with Marchand and his offer of a 'free passage' down the river; red, white and blue shirts, beads, umbrellas, etc.! Also a good story of the Dervish flag now hung in Les Invalides; not taken from a Dervish gunboat but from a store in Omdurman!

MARCH 13–15. Went in a boat to some small rocky islands below the fourth cataract and saw great numbers of geese. It was very hard towing up river for the stream is so rapid, but we sailed back easily all the way to Merowi. I found under a stone the nest, with five eggs, of the desert bullfinch; Charles says this nest has never been found before. In the evening Jackson Pasha visited us and told many interesting stories of Dervishes and slave-trading, which still goes on to a small extent.

MARCH 22. On camels from Merowi to Abu Hamed—about 120 miles. Sometimes the track lay close along the river, which is 'cataract' the whole way, and sometimes it went by an 'agaba', or short cut across a rocky part of the desert.... There is a constant change in the colours of the rocks and sand, hills and gulleys, and we were fortunate in having perfect weather the whole time—sleeping out every night without tents and in great comfort. There was a new moon on March 18 and I saw it for about half an hour before it set. I don't think I ever saw the moon before on the first day of its appearance. At Hebbeh we saw the wreck of Colonel Stewart's steamer. In '84 he and all his crew were murdered there by Dervishes. We have had some trouble with our camelmen—a lazy lot of scoundrels always wanting to stop—and this morning I had to kick them out of their sleep at 3.30 a.m. and follow them round to collect the camels. We were not sorry to arrive at Abu Hamed, and by great good luck Lord Frankfort—who had been half a day in front of us all the way from Merowi—was stopping here at an official house, so we called on him and got an excellent dinner and still more excellent lager beer.

Back at Cairo, they 'assumed the order of the boiled shirt and collar—very irksome after three months of flannel shirt and comfort'.

(*To his Sister.*) I am going East to catch bugs for N.C.R., and I hope to arrive in Japan about the beginning of June. When I have caught enough Japanese bugs I shall try to get taken on in some hospital as near to the war as possible—or go there in any other capacity—that is if they admit foreigners at all. Failing that, I don't know what I shall do. A globe-trotting invalid would perhaps be the best thing to be attached to but I doubt if I shall find one! Anyhow it is great luck to have a chance of going out to the East and if something turns up it will be all right, but this sort of hand-to-mouth business is not very satisfactory.

APRIL 15–23. (*Between Aden and Colombo.*) I have made the acquaintance of some of my fellow-passengers—a very pleasant lot and luckily not inclined for cricket and other steamship nuisances. Some of the lady passengers have come out in kimonos and look very alluring, but so far the only feat I have accomplished on this ship is to have fallen in love with Miss Eileen Fränckel, whose fifteenth birthday we celebrated on Monday. If only she had three or four of my years what a golden opportunity it would be. There was a dance on deck last night and I wished for the thousandth time I had acquired the art of dancing. At Colombo I smelt the real smell of the East for the first time and it was enough to make me wish for more.

MAY 4. Between 2 and 4 a.m. we came in for a series of most remarkable waves which made the ship roll prodigiously.

An angle of 38° was registered on the Bridge but we probably rolled at least 40°. Everything movable slid up and down the cabin, the rail of the main deck was several feet under water, and two stewards were nearly washed away in trying to close a door....Came into Fremantle, and a more one-horse forsaken town I was seldom in; three-quarters of the houses are 'bars' of some kind and the rest are hairdressers and tobacconists. In the morning the town looked as bad as it did by night.

Miss Eileen Fränckel tries to teach me the gentle art of dancing.

MAY 7. Two species of albatross to-day: one with black wings and back, white head, tail and under part, and the other much bigger with black wings, white back and flesh-coloured legs and bill. Tried unsuccessfully to catch them from the stern rail but the ship goes too fast.

Had some more lessons in dancing. Devilish temper all day.

On May 12 he is at Port Melbourne and walks out to the Observatory, where he is delighted to see numbers of laughing jackasses in the trees along the road.

Went to call on H. N. P. Wollaston, LL.D., Chief Controller of Customs, and a distant relation of mine. He was as deaf as a post, but very friendly, and he invited me to dinner, writing out very elaborate directions for finding his house—quite a Wollaston characteristic....I visited the Museum and Art Gallery where I saw a lot of rubbish but some good things too—especially a set of original Thames etchings by Whistler. In the evening I dined with Wollaston at his house and he and his family were most hospitable and professed to be very glad to see an English W., having never seen one before; it was a most jolly evening.

It may be because I have heard too much of Sydney Harbour, or it may be on account of the great cloud of smoke that hung over the city, that I was not so very greatly impressed. I think Falmouth is almost as fine; Fowey is infinitely prettier. I wandered about the town and thought it a beastly place with beastly dirty shoddy buildings—tramcars making such hideous noises that I had to take refuge in the Botanical Gardens. I put up at the Australia Hotel, but went back to lunch on board the ship, for I felt a good deal more at home among the German-speaking officers on board than at the hotel among Australian-speaking natives. Visited the N.S.W. National Art Gallery and amongst a great deal of fearful rubbish saw some good etchings of Axel Haig and a few nice sepia sketches of Prout; and of all things in the world E. H. Fahey's picture of the old canal or stream at Kingswood near Wotton. In the picture gallery at Adelaide I had seen Fahey's picture, 'I'm going a-milking, sir, she said', being copied by an elderly student. Fahey would be flattered to know that his was the only picture being copied in the whole collection. The average pictures in these galleries are mostly villainous copies of old masters.

MAY 22. Celebrated my birthday with Sidney McDougall (of King's) in a real Australian way. We went by ferry and tram to a place called the Spit, in Middle Harbour. There we got provisions and hired a light double-sculling boat which we rowed a couple of miles up one of the many creeks. We camped on a bank and made a fire. The smell of burning gum leaves was new to me and I shall never forget it. We boiled the 'billy', made tea in it, and consumed a tinned tongue, bread and apples. It rained most of the time, but we had brought a sail with us which made a good shelter and I at all events was supremely happy. We might have been Captain Cooks for all the sounds and signs of man that were there, though we were not more than 3 miles from a big city. When the rain stopped we wandered up through real 'bush'—the first I had ever been in—great gum trees of all shades of green and brown, some with naked trunks and some peeling, and an undergrowth of many flowering shrubs. There was only one thing lacking and that was birds: I saw just two and they were silent.

On May 25 he leaves Australia, and coming into smooth waters sees the sun rise over the outer islands of New Zealand.

JUNE 1. These islands are not unlike those along the Norwegian coast but the mainland is lower and thickly wooded. I landed at Auckland and went for a 5-mile walk inland. Found myself in very English-looking fields—hawthorn hedges, oaks, Scotch firs and poplars; the air full of the songs of thrushes and blackbirds, larks and goldfinches, so that I could scarcely believe that I was not at home. Went up a small hill which was an old volcano with a perfect little crater on the top, and it was cold—much colder than Sydney—and I had to walk hard all the way back to keep warm. I wandered about Auckland and was astonished at the number of good bookshops—at least half a dozen. In Sydney I could only find two and in Melbourne one.

Between Auckland and Rotorna the country was somewhat like Westmorland—small rocky fields with stone walls; later it became wide marshes—a splendid hunting-ground for water birds and bugs. Walking back from Whakarewarewa, hearing the birds singing and feeling the chill cold air, I could well imagine myself walking in England on a late autumn evening.

JUNE 20. Went to Rotoiti, and in the little channel between Lake Rotorna and Lake Rotoiti there were beautiful deep pools and eddies with shelving banks. A better place for fish I never saw. Shoals of trout, great big fellows—ten pounders at least. I would have given a great deal to have had a rod in my hand.

Travelled to Wellington, which is a dirty little town, or rather 'city', for all towns are cities here. I am struck by the number of loafers. It was the same at Auckland, Napier and

Sydney: at every corner of the streets and at all hours of the day you see groups of young men from sixteen to thirty doing nothing—quite decently dressed but with nothing to do.

In the museum at Christchurch I saw a splendid collection of moas and extinct New Zealand birds. The curator is one F. Wollaston Hutton, F.R.S., so I had the audacity to call on him. He is a most interesting and genial man, not the least bit of a fossil, and has just published a popular book on the animals and birds of New Zealand which I must get. He tells me that the great finds of moa bones were a very gold mine to the Museum. They not only kept a complete collection for themselves but managed to send some to almost every other museum in the world, receiving in exchange all sorts of valuable things—notably a good collection of Greek and Roman antiquities which one hardly expects to see in New Zealand.

JUNE 21–28. Sailed from Bluff and steamed round the Tasman peninsula and Cape Raoul, with its wonderful basaltic columns like the Giant's Causeway. When I got to Hobart I walked as far up the big hill immediately behind the town as I could get without being in the clouds. I came to the conclusion that the blue gum tree is very much maligned: it has a magnificent trunk and branches and the light and colours of the bark are often beautiful. I saw a flock of big black cockatoos with yellow crests; they tear off the bark of the gum trees and leave it hanging in great festoons.

Here I went to dine with a delightful family called Giblin (an old Kingsman). They are all musical and the lady sang Schubert and Schumann and played Beethoven most of the evening, so I was entirely and completely happy. I was obliged to contribute 'Widdicombe Fair'.

JULY 6–18. Left Australia, and on July 14 sailed into the bay of Herbertshöhe, New Britain.

I went a few miles across the bay to Matapi and walked through some native villages and up through a coconut plantation into fields of tapioca and yams, and then got into uncleaned bush where I had plenty to do in catching butterflies for N.C.R. There are any amount of bugs about, but it is hard work catching them as the jungle is tremendously thick. The grass 5–6 feet high, ferns, creeping convolvulus, vines and pumpkins, with a mass of creepers hanging down from the trees, all this makes any sort of progress desperately slow. I caught about twenty butterflies and returned to the ship hotter and in a greater state of sweat than ever before.

At Simpsons Havn I had a fine scramble through jungle where I caught a good many butterflies, but the best were quite inaccessible. I watched a beautiful *Ornithoptera* sunning himself on a high tree and wished I could catch him. Came down a dried watercourse that was simply alive with insects and beautiful plants, but no orchids growing on the trees as I had hoped. Made the acquaintance of 'Sour sap'. It is a large green prickly fruit with a white creamy inside which you squeeze through a cloth and drink with a dash of claret —very good indeed.

JULY 19. Left New Britain and sailed round north-east end of island keeping pretty close inshore. The country from the ship looked absolutely wild save for one or two plantations, and I was told that if you walk a few miles from any house it may mean a knock on the head for you and a square meal for the natives.

This morning I saw ranges of big mountains to the south-east (Finisterre) and to the south (Bismarck). Later we came through a narrow channel between a small island and the mainland and dropped anchor a few hundred yards from the wharf of Friedrich Wilhelm's Haven. Went ashore and at the edge of the jungle I set about catching butterflies. It was blazing hot and I was melting in no time, but caught about fifty of quite thirty species including two *Ornithopteras* in fairly good condition. Unfortunately I was too much occupied to look out for birds.

JULY 21. Rowed over to an island. The sea was perfectly smooth and I looked down on sponges and lovely pink and white and yellow corals. On the island beach there were a number of men busy building canoes, and as I was in the act of taking photographs two men armed with long spears and big painted shields rushed out from behind a house and pretended to attack me. It nearly scared me out of my life but the whole village enjoyed the joke immensely. I got a small boy to help me catch butterflies with my net; he killed them very neatly and brought them back saying, 'He die finish', i.e. dead, for 'he die' alone means he is sick. All communications between whites and natives in these German colonies is carried on in pidgin English—never a word of German. Even on board ship the officers have to use English to the native sailors and firemen, and I have seen black men being drilled and taught to march to the tune of Left Right, Left Right. At Matapi, when we asked some boys from a mission school if they were Germans, they answered, 'No damn fear, me English'. Some of the pidgin English expressions are rather comic. For instance at a piano recital I went to, the instrument was called, 'Big fellow box, man fight him he cry plenty too much'. A saw is described as, 'He come, he go'.

The crew of a Layland sailing ship that was wrecked some days ago on a reef about 300 miles east have just come in here; the boats are still missing, and their chances I understand are not particularly good as the natives are all cannibals right away down there.

JULY 26. Along the New Guinea coast for the last few days, but now we have turned a bit south through the Pitt Strait—about 2 miles wide and 25 miles long. It runs between two islands, on one of which some few years ago a party of officers went ashore and were killed and eaten by natives.

JULY 27. Came to the beautiful little harbour and town of Banda, where the houses are built out into the water on piles and are low and single-storied with broad palm-thatched roofs and wide verandahs; no corrugated iron in the place. On a hill behind the town is an old Portuguese fort, early seventeenth century, and the remains of another with a wide moat and Anno 1617 carved on the wall. Here I landed for the first time in a Dutch port and in a Malay town. We have left the country of Kanakas and naked 'savages' and have come into what seems, by comparison, the height of civilization. The Malays in 'sarongs' and some even in shirts appear almost indecently overdressed. Most of the traders and shopkeepers are Chinamen. I walked up into a wood of nutmeg trees now in full fruit. I cannot make out whether they are native here; some are certainly planted, but the greater number grow haphazard. I managed to catch a few butterflies and then walked down into the market where I found a large crowd round the body of an alligator which had just been killed in the harbour. It was carried about in a sort of triumphal procession from one part of the town to another so that everyone might have a chance of seeing it. This is a great place for parrots, and many of the natives carry three or four on a stick to sell—some of them most beautiful birds.

JULY 29. From Amboina we sailed due west and dropped anchor off the town of Macassar. Here the houses are made of brick and plaster and are more or less of the Dutch type, but as we walked into the outskirts of the town, the houses became more Malay in character, and a great many were real Macassar houses, built of bamboo and nothing else. They look as if a puff of wind would send them down like a pack of cards, but perhaps winds are rare or the houses stronger than they look. It is without exception the most evil-smelling place I have been in. An open drain runs down each side of the streets and every shop sells quantities of raw fish only very partially dried. Drove out to call on the Rajah of Goa. Three miles from Macassar, Dutch authority ceases and this potentate reigns supreme. He has powers of life and death and is reported to have immense wealth which he hides in the earth in different parts of his domain. The coolies who do the hiding for him are killed when they have done their work. He sat in a great square tin-roofed building like a barn on two floors. He looked a most desperate old scoundrel dressed in a dirty white jacket, round cap, sarong and bare feet; his teeth black from betel nut and his cleanliness very much below par. The only pictures in his state chamber were an advertisement of somebody's beer and another of somebody's cognac. He produced some indifferent cigars and examined my field-glass case. Unluckily I had left the glasses behind, for I might have changed them for some of his big pearls. About a quarter of an hour was enough, and I came away through the market place of Goa where gambling and cockfighting were going on amidst great excitement. The crowd was worth going some miles to see.

On August 3 A.F.R. comes to Java, and travels about in a country of broad rivers and wide plains with a background of splendid mountains. He finds a snug little hotel at Garoet and catches a death's head moth in his bedroom, 2200 feet above the sea.

AUGUST 11. Off in a cart drawn by three ponies for Papandajan, a volcano about 18 miles away. On the way I saw two huge flying foxes going home to bed; they looked bigger than the biggest owl and fly very straightly and deliberately. Up through coffee and quinine plantations into real Java forest, and it was interesting to watch the change in the vegetation from 3000 to a bit more than 6000 feet: at first pure tropical jungle, then tree fern zone, then smaller ferns, and lastly scrubby stuff near the crater. The volcano is not active now, but there are plenty of steam holes, solfataras and fumaroles.

At Djokjakarta I watched women making sarongs. The designs are all put on by hand, sometimes following a tracing held behind the cloth and seen when it is held up to the light, and at other times without any tracing at all. The colour is mixed with wax and is laid on with a fine tube coming out of a little brass cup. Nearly all the sarongs and head dresses here are blue and therefore not nearly so varied and gay as those in the other places I have been to.

I went to see the Buddhist temple at Boroborda, but more pleasing to the eye than the unending carvings of Buddha was the grand view of volcanoes, peaceful mountains and fertile plains. I am much disappointed in the teak plantations, for I had always imagined the teak tree to have a vast big trunk, but they are poor things, with hardly 10 feet of the straight to them.

At Batavia I went to the Zoological and Botanical Gardens. To my disgust they were full of drinking bars and bandstands. Not a bird or beast in the place and certainly not more than a dozen kinds of trees—none of them labelled. After sunset I saw fifty or more flying foxes all starting out for the night, all going in the same direction, and looking as they flapped along—singly and slowly—like rooks going home. In the Botanical Gardens at Singapore there were a lot of good trees, but they are still young—not more than 15–20 years old.

Tiffin here vile, only relieved by mangostenes, a really excellent fruit with a very elusive kind of taste which you just begin to realize when you have finished the flesh and have to eject the stone; decidedly the best tropical fruit I have met.

Sailed for Japan, and soon after daybreak on September 4 we came into Nagasaki Harbour. Outside the harbour a small steamer flying a war flag met us and led us carefully up the harbour to keep clear of the mines with which the place is plentifully strewn. I passed a very particular medical examination conducted by five Japanese doctors and was eventually allowed to land. The place was full of great celebrations of the Japanese victory of Liao Yang. The whole population was out dressed in its best and everyone carried either a paper lantern or a flag. The colours of the decorations were confined to red and white (the Japanese colours) and the result was unquestionably better than anything in the way of street decorations in England.

Much to my disappointment no Russian warships attempted to stop us, and so far, I have not succeeded in meeting any man big enough to give me a pass into Korea, so I begin to fear I shall not get there.

SEPTEMBER 10. Left Nagasaki and went 60 miles through hilly country and then along the seashore to Arita—famous for its potteries. I believe I could watch all day a man take a big lump of clay, spin the wheel round with an occasional touch of his foot, and then—as by magic—evolve from the top of it a beautifully shaped Saki flask. Painting the designs is pretty work too, but there is a good deal of tracing paper used and it is more mechanical.

SEPTEMBER 11. This native hotel at Kumamoto is very comfortable, but I shall soon tire of doing without chairs and tables. My back and all my joints ache in a way they never did before, and I find even writing in this notebook a fearful labour, performed now in one position and after a line or two in another. Four maidens have just brought me tea and helped me to drink it with much conversation and laughter begotten of the 'conversation sentences' in my little handbook. They are quite engaging in their ways—after a sort of kittenish style—but to my mind they are more ugly than pretty.

SEPTEMBER 13. Left Kumamoto in a rickshaw pulled by two men. On our road we met streams of country people all carrying flags and dressed in their best, which puzzled me until we came to Miyaji, where I discovered that the news of the victory of Liao Yang had only just reached them—such is the remoteness of the village. The people were certainly not accustomed to foreigners for I was surrounded and stared at more like a wild beast than I have ever been before, which is saying a good deal. My guide says it is my fair hair which attracts so much attention; anyhow it is all most embarrassing. In an open place where the crowd was thickest I noticed a square platform, in a corner of which a man and boy were playing a dreary tune on a drum and penny whistle. I was told that we were just in time to see a Nō dance. Ten men advanced slowly up an inclined plane on to the platform; two of them had no masks, and one had a mask rather like a woman; the other seven had masks supposed to represent the god Nō, and in that case he was the most hideous deity that ever came from heaven. Some wore wigs of long red tow, others of white, and they all wore short kimonos beautifully embroidered. To me their dance was grotesque and almost repulsive. If I had not been a sympathizer with Russia before I saw it I should certainly have been so afterwards. Later on I was entertained by an amateur theatrical troupe from another village. Unluckily I could not understand a word, but I could appreciate the excellence of the acting, which was uncommonly good: the tones, gestures and attitudes were often quite admirable—and these were common people from a small country village. The big low-ceiled room lit by a couple of lamps, the crowds of delighted children sitting on one side and the equally delighted guests of the inn sitting on the other, with this extraordinary display going on in the middle, made a scene which I think I shall always remember.

No signs of festivity this morning, and as we left the

village the children were trotting off to school as if the Nō dance had died a thousand years ago and Liao Yang had never been captured....We climbed about 1000 feet up a steep road and the sun blazed hotter and hotter with every mile that we went. Wild flowers grew in masses by the roadside. The balsams are, I think, the same kind as the 'touch-me-not' that grows in gardens at home, and there was a lovely pink—like the large cheddar pink—a very pretty white clematis, and heaps more whose names I know not. They cultivate paulownias here and use the wood for making shoes.

SEPTEMBER 15. Left Takeda by a tunnel through hills to the north-east. It was much cooler, which pleased the rickshaw men very much, and we bowled steadily downhill with occasional glimpses of the sea to the east until we came to Oita. Rain makes a great difference to the consumption of rickshaw coolies' shoes. In good weather they only use two or three pairs a day, but in wet weather the score goes up to five. It is not so bad as it sounds, as the price of a pair of grass-rope shoes is only two yen—about a halfpenny. I have discovered that a tree which has puzzled me a good deal is the wax tree. The small round berries, rather like big elderberries, are boiled down and the wax which is extracted from them used for candles.

SEPTEMBER 17. (*Beppu.*) This evening I went out on to the beach and into a hot bath-house where men, women and children bathers—all stark naked—stroll about in the most promiscuous way. While I was having my bath a lady came in in the most matter-of-fact way in the world, and although I was actually in the bath she sat on the edge and began to wash herself. My sense of propriety and modesty overcame me and I beat a hasty retreat. I am surprised to see so many of these people suffering from skin diseases, face eruptions and such like. It is hard to reconcile this with the large amount of bathing done, but I daresay the insanitary surroundings of the houses account for it. There is plenty of room here for an aspiring dermatologist and an oculist too, for the number of blind and one-eyed people is almost as great as in Egypt.

SEPTEMBER 19. (*Moji.*) Took a sampan to Shimonoseki. Here I found myself back in a European hotel, with chairs, tablecloths, bedspreads and other impediments. It all seems fearfully stuffy and over-furnished after the artistic emptiness of the rooms in a Japanese inn, and I almost regret the latter —in spite of aching joints. I was told to-day that the art of arranging flowers which I have admired so much in all Japanese houses down to the very meanest—and which I thought was their natural good taste—is carefully taught to the people. In some parts of the country ladies spend a great deal of their time in teaching the people how to do things prettily. I always thought Father could arrange flowers better than anyone, but I rather think a good Japanese will do as well.

SEPTEMBER 21. (*Miyajima.*) Went out intending to go for a walk but I found it impossible to go far, for every yard there was something to stop and look at—a pretty house, or temple, or tree, or bit of sea. There is a Japanese proverb: 'Call nothing beautiful until you have seen Miyajima', and I am disposed to think it is right, for I was never in so enchanting a place in my life and I could go on raving about it for pages. The island is holy on account of its famous Shinto Temple, and no carts, rickshaws or dogs are allowed on it, nor are births or burials, so it is the most peaceful spot on earth. The wild deer of the island are so tame that they come down to the shore and take food from your hand, and there are always some feeding in the temple grounds, making a very pretty picture with the children and the pigeons. A smiling old woman gave me a little plate of corn, then she rang a bell and the pigeons came in clouds from a big dovecote and settled all over me so that I was quite smothered in pigeons. She had a white stork and two big cranes in a little garden about 4 yards square, and they looked much more contented than they do in the big paddocks in the Zoo at home. The pines are one of the greatest joys of the place, being of such a glorious colour and growing in beautiful forms, curved and straight and twisted, and always in the right place; the people love them, as can be seen from the way they prop up a failing branch rather than cut it or let it be blown down. There are some fine old cryptomerias near my inn and a great many maples which will be glorious in a month's time.

SEPTEMBER 22. Have been given an introduction to an American lady doctor who is superintending the nursing in the hospitals at Hiroshima. This is just what I wanted, and so I took a sampan over the straits soon after sunrise and found Hiroshima swarming with soldiers and horses and wagons, with sentries at every street corner and outside many of the houses. I went through many barracks and stables, past parties of buglers learning to march and bugle at the same time, until I found the big reserve hospital. It was full to overflowing. Hundreds of the less severe cases were housed in temples and sheds, and they are building new hospitals as fast as ever they can. Dr McGee welcomed me most kindly and gave me her whole morning from ten to one o'clock, and although I have never met a woman doctor before and am not a little prejudiced against them I shall certainly have to alter my opinions a bit. She told me a lot of interesting things about the war and about the pluck of the wounded soldiers. The Japanese system of nursing is utterly different from ours, where the nurse is boss and the patient does what he is told. Here the patient tells the nurse what he wants, asks for his food at odd times if he wants it, and even takes his own medicine. One would have thought this would have been a dangerous plan but on the whole it seems to work remarkably well. Of all the wards that I went into the only one that was still and quiet like the ward in an English hospital was that where the very worst cases were lying, wretched men who had been shot in the spine, who cannot move and will probably never move again; the other wards were the most cheerful places imaginable. In one I saw a hundred men just beginning their midday meal, and really it was more like a school-treat than anything else. They were laughing and talking and enjoying themselves immensely. One fellow with three severe wounds and just out of the operating room was in bed smoking a pipe and writing a postcard; another, with a hole through his head, his shoulder and his belly, gave me a graphic description of how the bullets came (with appropriate sounds), and thought it all one of the finest jokes in the world. This is typical of the whole lot; I didn't hear a cry or a groan or complaint, and the nurses tell me they never do complain. I think they really feel pain every bit as much as European people but it is a point of honour among them to hide their feelings.

Much to A.F.R.'s regret he failed in his efforts to get taken on as a doctor near to the seat of war, and so he prepared to leave Japan. A few days later he sailed from Yokohama and on October 15 he arrived at Vancouver. He went to visit his brother Francis on his ranch at Kelowna, and here had what he considered one of the best times of his life.

What I enjoyed most was acquiring some of the art of riding. The 'cayuse' of the country is not very big, a bit over fourteen hands and fairly docile. Every Saturday paperchases took place, when about thirty people including myself galloped over the country. I must confess to having been a bit nervous at my first appearance on horseback, but no ill befell me and I enjoyed it vastly. My animal 'Socks' always kept up with all but the very best of them and never got me into trouble. We had one or two days shooting duck and prairie chicken but never got very many birds. There are heaps of nice people here, all most hospitable, and I think in one month I went into more houses than I go into in London in one year.

On December 10 I arrived back in England, and travelling to London went straight to a Richter concert at the Queen's Hall.

It is characteristic of A.F.R. that the moment he arrived home he should go straight off to hear music. He used often to say that after months spent in out-of-the-way places a great craving for music would come over him, and that one of the first things he would do on getting back to civilization was to go to a concert. He was no musician himself, but his mother was an extremely fine pianist. No one who ever heard her play can forget the pleasure her music gave, and undoubtedly it was her constant playing at home that enabled him as a boy to hear and love all that is best in music. Whilst up at King's he would go to a concert whenever it was possible. Once after hearing Joachim and Leonard Borwick play he said: 'I think I never enjoyed music so much before. It makes me mad to think what a fool I was to give up playing the fiddle when I was a boy. Who knows, I might by this time have played quite respectably, and I haven't the courage now to begin all over again'.

In 1906, on his return from the Ruwenzori Mountains to a little hotel at Entebbe, he wrote: 'I listened this evening to a little German clerk playing on a shockingly out-of-tune piano. It was the first music I had heard for ten weary months. I made him keep it up till midnight, when his arms and hands were worn out. He had not much execution, but the true spirit of Beethoven, which put me into such an "ecstasy of joy", to quote Mr Pepys, that I could not sleep. Music is the one indispensable luxury—no, it is a necessity'.

V

RUWENZORI LETTERS

On October 10, 1905, A.F.R. was appointed Assistant House Surgeon at Addenbrooke's Hospital, Cambridge. Two days afterwards, when he 'had just settled down to the awful prospect of life in a hospital', he heard of the Ruwenzori Expedition wanting a doctor.

(*Diary.*) I wrote to Ogilvie-Grant at the British Museum, saw him in London at the B.O.C. dinner on the 18th, and fixed it up. I had some difficulties with the hospital authorities, but found a substitute and got away from there on the 26th of October with only a fortnight to make my preparations. The rest of the expedition had left Genoa some ten days earlier. It seems that on this expedition my *raison d'être* is in one of the fellows on the expedition—G. Legge, a son of Lord Dartmouth. His parents do not like the notion of his being in distant parts without some medical assistance available so I go as doctor and botanist, and am to collect and share in the very problematical profits of the expedition and am paid £12 a month.

This was the British Museum Expedition to Ruwenzori, under the leadership of R. B. Woosnam. The other members of the party were R. E. Dent, The Hon. G. Legge, Douglas Carruthers and A.F.R. The primary object of the expedition was to collect specimens of the fauna and flora of the district, and though a good deal of climbing in rough places had to be done by the collectors no special mountaineering ascents were considered probable. One or two of the party, however, found it impossible to leave Ruwenzori without attempting some of the peaks, and it was on this expedition that A.F.R., Woosnam and Carruthers made the first ascent of what was then supposed to be the highest peak in Ruwenzori, now named Wollaston Peak, 15,286 feet. In September 1906 the expedition came to an end, but A.F.R. and Carruthers decided to stay out for some months longer and travel home through the Congo. Their journey across Africa, together with an account of the time spent in Ruwenzori, is described in *From Ruwenzori to the Congo*,* which A.F.R. wrote when he came home. In the writing of this book A.F.R. drew almost entirely from his diaries, so I have decided not to publish these, but only some of the letters he wrote home at the time.

DECEMBER 23, 1905. (*Entebbe. To his Father.*) I have been kicking my heels in the so-called hotel here waiting for my baggage to arrive by the weekly steamer, and now that it has come I am busy sorting it out and getting it into 60 lb. loads. There are a hundred and one things that have to be remembered and it is particularly trying in this hot rainy weather, but to-morrow morning I hope all will be well and I shall start off west. My party consists of twenty-six porters, headman (not an executioner, but boss of the porters), cook, 'boy' and myself. It is about fifteen or sixteen days' walk from here to Toro (Fort Portal) and from there I expect two or three days' from the first camp of the rest of the party. I shall have rather a lonesome kind of Christmas—the first few days are through a very wet and swampy country—and I could wish for a companion on the journey, but I am taking a bottle of Lager beer and intend to drink your healths in it, as I hope you will do mine. Entebbe is not a bad sort of place but I shall be glad to get out of it into barbarism. Another occupation has been struggling with the Swahili language—a frightful collection of incoherent sounds. So far, I have attained to the proud position of being able to call a man the offspring of a serpent and a few other compliments, but in time I hope my conversation will become less personal. There is great competition now to make the first ascent of Ruwenzori. Douglas Freshfield and Mumm came back disappointed a fortnight ago, and yesterday an Australian called Grauer started from here to

* Published by John Murray, 1908.

make an attempt. If he fails he is going to have another try with me in February or March, when I hope we shall have luck. By all accounts the weather is so bad there that one does not like to be too hopeful. If I don't go and get that tin-opener now I am altogether lost!

JANUARY 16, 1906. (*Bihunga. To his Father.*) Here at last. A journey ten times as long wouldn't have been too much to get to such a glorious place. The thirteen days to Toro I confess were rather monotonous, up and down hill with hardly ever a view of the country. Here we are about 7000 feet up, perched on a spur of one of the ridges overlooking the valley. We live, work and eat, and some of us sleep, in a fine grass house built before I came up. The rest of the party are in tents. The four men seem to be a very good sort and I think I shall get on well with them. Funnily enough I struck up an acquaintance with one of them (Dent) in a train last summer. He told me he was going to Uganda, and when we got to Euston I said to him 'perhaps we shall meet in the Semliki forest', though of course I had not the least notion that I was going to do anything more exciting than 'public health' at Cambridge.... We are just at the upper limit of bananas here, about the level of dracaenas and giant groundsels. Above is big forest, and then bamboos up to 9000 feet which is the furthest I have been so far. I forget whether I told you that in spite of my complete ignorance of the subject I am botanist to the expedition. I make expeditions armed with a botanical tin and a butterfly net and bring back large captures, botanical and entomological. My pressed plants are beginning to look very nice, but I fear for them when the wet season comes, as it will before long. Even now, the driest time of the year, everything is very damp....

FEBRUARY 26. (*To J. H. Clapham.*) I mean to come home by the Congo and hope to be in England about the middle of next year. After that, well I don't know. I am getting seriously alarmed about myself. Often I wish that I had never been out of England in my life, for the more I wander about the less inclined I am to stay at home. I thought a year ago that I had come back for good, but at the first possible chance I threw up the work I had and was off here. I have a notion that unless a man has fixed himself in his round or square hole by the time he is five-and-thirty at the latest he will never do so, and I have not much time left. The worst is that I have no money and this sort of thing does not pay. I cannot write, as you always imagine I can. It is true that I am always writing something and it goes well for a time, but then I get stumped for a word or expression which won't come and there's an end of it. I wrote rather a jolly thing about Naivasha as I was walking along the road from Entebbe. It went as smoothly as possible until I came to a great expanse of coarse grass where the Masai feed their sheep and the wild gazelles feed with them. Something happened, and I couldn't fit them in properly, and there the thing is—stuck fast.

APRIL 23. (*Muhokya. To his Father.*) We have now come out of the mountains into the plains. After being cooped up in a narrow valley with huge mountains and forests all round, it was like coming out of prison to see a wide horizon and level ground in front of you. Here we are three days' journey down the east side of Ruwenzori in the direction of Lake Albert Edward. We are camped on rising ground with the lower spurs of Ruwenzori behind us, and in front a brilliant plain of grass, sedges and acacias; a mile or two to Lake Ruisamba, and beyond that the purple hills of Ankole. It is a beautiful place to camp in, but there are drawbacks, and here they take the form of ticks and mosquitoes and flies and heat. Nobody has had fever yet but they are none of them so tough against heat as I had hopes they might be, and we shall have trouble no doubt before long. I don't think I have told you that we made another expedition to the snows and climbed the peak called Duwoni* which Freshfield and others thought was the highest in the range, but it is less than the other peak Kiyanja† which we climbed to within 150 feet of the top. The really highest peak can, I believe, only be reached from the Semliki valley side, so we are hoping to get round there before the Duke of the Abruzzi comes. At present we are absolutely stuck, as the British Museum will not send funds, and without them we cannot move. Two days ago Woosnam and I walked down to Lake Albert Edward across the boundary into the Congo Free State to get information about the other side of the mountains. The boundary is a river which was in flood and not too safe, but an officer in charge of the post came down to meet us, and with the help of a tatterdemalion collection of niggers, formed a rope of odds and ends of box-cord with which we passed some of our clothes and boots across to the other side of the water. I tried to get across in the same way, but after a yard or two the rope broke and I was swung down by the current. Luckily the blacks at my end hung on tight or I should have gone sailing down to the lake at 20 miles an hour. After that I thought it better to swim across the big pool higher up where the current was not quite so swift, though swift enough for me. I was clothed in a sun helmet, and never had I crossed a frontier in such fashion before. The officer was a very nice fellow and made us welcome even to the extent of surrendering his easy chair and the only bed; a very hard floor fell to my lot.... To save us the bother of swimming across the river, this officer next morning took us down to the lake and gave us a dugout canoe to go to Katwe. But this proved to be a much longer way, and we did not get back to camp till late in the evening—both pretty well done up.

* The twin peaks of this mountain are now called Wollaston Peak, 15,286 feet, and Moore Peak, 15,269 feet.

† Now called King Edward Peak, 15,988 feet. A.F.R. and Woosnam climbed to a height of 15,840 feet, and their ascent of this mountain was higher by several hundred feet than that of any of their predecessors in Ruwenzori.

MAY 15. (*Muhokya. To J. H. Clapham.*) ...From my tent I look down a slope of scattered thorn trees to a wide shamba (banana plantation) which looks pleasant and cool with smooth green leaves, then a broad plain with belts of trees and baking grass, then Ruisamba, and beyond that the Ankole hills—generally purple, but this morning faint with haze; white clouds and hot sky—a prospect fair enough indeed, but the rose has prickles. Here they take the forms of fever-producing ticks by day and mosquitoes by night. There is plenty to do, bugs to catch, plants to collect, and big game to stalk (with a field-glass)—but my heart's in the mountains and I am consumed with desire to be up and doing. We ought very soon to be starting round to the Semliki side of the range but we are stuck here by reason of exhausted funds and can move neither forwards nor backwards. Meantime the Duke of the Abruzzi draws near with a great following and will be at the top of the highest peak before I get within 50 miles of its foot. It is one of the most grievous ill chances that ever befell me and I am inclined to curse all royal dukes and wish they would stop at home. But I met with a certain measure of success up there. I think I told you of our ascent of Kiyanja. I went up again to measure the height more accurately and found it to be a good deal higher than Duwoni, the two-topped peak which Freshfield thought was the highest....Stuhlmann photographed from the west a very big peak with two tops. Freshfield saw from the east a big peak with two tops, and jumped to the conclusion that they were one and the same thing. I went to within 150 feet of the top of Kiyanja and saw both. Stuhlmann's is much higher and a long way to the north-west of Freshfield's (Duwoni). The high peak is what has been called Saddle Mountain.... Johnston's estimate of the amount of glaciers is of course very much exaggerated, but there are some other big ones in a hidden valley on the west side of Kiyanja and doubtless in other places not yet visited.

I hope all this doesn't bore you. I often walk out on to

the plain and tear my hair when I look up at all these peaks and valleys that no one has even peeped into. The D. of the A. will have his work cut out for him to get to the highest peak from the Mubuku but no doubt he brings European porters. If he can get a tent on to the upper snows all will be plain sailing. I mean to get round to the west side somehow and at all events get a view of the peaks from that side.... Just think of going over from Tanganyika into the Congo! Never mind, you have as you say your reward, and you have an English spring now and strawberries by the time this reaches you—whilst I am a rotten old ne'erdoweel, starving on goat and bananas. (Heavens, how sick of them I am!)....I wonder what Burton would have done with the languages here. I started on Swahili, but that is no good here; you want smatterings of Luganda, Lunyoro, Toro, Lukonjo and a dozen others. I have given it up, and trust to my eyes rather than to my ears. Dauphiné sounds jolly; I must try and get there next year with, I suppose, a bachelor companion. Take off your hat to the Alps for me and let me hear more at your earliest pleasure. I wish I were a Bateleur Eagle;* there is one flying over the camp now. Hawthorn and dog roses—I would give something for five minutes of them now. It's the only time of year that I get homesick. Lucky this is the end of the sheet or I might become sentimental.

MAY 20. (*Mubokya. To his Father.*) I am longing to get to the west side of the mountains from the Semliki valley, but we are compelled to mark time here by reason of a lack (not a lakh) of rupees which the B.M. cannot find or cannot send....At present there is no one who knows the real relation of these peaks, and it really is rather hard, after getting to the top (in the wet season) of the two supposed highest

* In his diary he says: 'For power and swiftness of flight there is none that can compare with the Bateleur Eagle. Its long wings and curious stunted tail make it look more like a huge bat than a bird as it sails high overhead—never flapping its wings, but just giving an occasional tilt from one side to the other. At one moment it is here—and the next it is a speck almost out of sight'.

peaks, to leave the really highest for another to climb in fine weather and with all the accompaniments of guides and porters and other paraphernalia.

Did I tell you of my projected route homewards? If nothing happens to prevent it I mean to come back through Ankole to the north end of Albert Edward, down the lake in canoes to Lake Kivu and so to Tanganyika. Then three days' march down the west side, meeting the Congo at a place called Kasongo: thence in canoes to Stanley Falls and after that by steamer and railway....It is possible that Carruthers may come too, but I shall go that way in any case, whether he comes or not.

There is a small round lake, an old crater, a few miles from here, and beloved of hippos. It is a ludicrous sight to see them bathing in parties of twenty or so. They bob up and down within a few yards of the shore and as the water deepens very rapidly they cannot go far out. They submerge quietly and come up with a tremendous snort, and yawn prodigiously. At night they make an infernal grunting. The same lake is full of fish which die for some unknown reason in extraordinary numbers. The shallows are full of half-dead fish and along the shore is a pile of bones—in some places a foot high. There are a number of old craters about but no signs of recent volcanicity (if there is such a word!). I have made a large collection of butterflies and have a regular farm of caterpillars, but the flowering plants are not so many as I had hoped. Some of the white acacias here are just coming into flower and look at a distance very much like hawthorn. A sight and smell of hawthorn now would do one a power of good. I am very glad to have acquired a taste for plants but I wish I could do it in a more intelligent way. I know no more now about African plants, having no books, than I did five months ago. You ask me about vegetables. Well, there is one thing—and it must belong to the family of Arums—whose root is excellent boiled or roasted, and whose leaves (rather like shiny elephant's ears) make most excellent spinach. I am afraid it would not do in England though it might in the Channel Islands or Scilly. I am not allowed to collect things for myself but I am sure there would be no objection to my keeping some seeds. I have a good number of different seeds, chiefly from up high, which ought to do well in the south-west of England if I can get them home all right; but the damp is playing the very deuce with them as with my pressed specimens. I am hoping to hear soon that my first batch of dead and living arrived home safely. With any luck I ought to get some good plants on my way back by the Congo; a few valuable orchids would not come amiss.

JUNE 1. (*Fort Portal. To his Father.*) You will wonder at this address, but I have come over here to meet the Duke of the Abruzzi and give him all possible information about the mountains. The Duke has with him Sella—the great photographer of mountains—and a geologist and doctor, two Courmayeur guides and four Italian-Swiss porters: altogether the greatest crowd of Europeans that was ever seen here. The Duke is very pleasant and easy to get on with, and I have told him all I can about the peaks, for which he seems to be grateful. He is lean and tough, about forty I should think, and a climber by the look of him. He knows Collie and many more English climbers, so there is always plenty to talk about. With their splendid equipment and European porters they should have no difficulty in climbing all the big peaks. They start to-day for the Mubuku valley, and as our roads are the same to begin with I go with them for the first day. Sella, as photographer, will illustrate the great book they are going to write; it ought to be worth having. It is very interesting to meet all these people, but I must say I wish they had not come this year, as I should dearly like to have been first on the highest peak; but with two good peaks to my credit I cannot complain....I sent a short notice of our climbs to the Secretary of the British Museum to be published if he thought fit.

I wonder if you will see it in any of the papers.... We are still stranded for want of money and likely to remain so. It was all I could do to raise ten rupees for porters to come here and back, although travelling is not very expensive. My five porters carry 50–60 lb. each on their heads for 60 miles and back again, and all for about two shillings and twopence. A chicken (very small, it is true) costs a penny farthing, and eggs (mostly rotten) a hundred for a rupee. Then I have a couple of loaves of bread, a tin of cabin biscuits, one or two tins of sardines and a tin of jam. In the Congo you pay porters in calico (called Americani) and beads—rather a nuisance as it adds considerably to your baggage. Well, so much for me. Talking to these two European guides has given me a terrible craving for the Alps. They know the Gaspoz and many guides that I know, so we talked mountains yesterday for an hour or more. Next July, if all goes well, I hope to go there again, but I rather doubt if I shall see much of my hard-earned £12 a month. When I get back I hope to make a few pounds by selling some of my negatives.

My salutations to the Mendips. I am longing to see them again, and Dundry and the view down Channel....

JULY 4. (*Muhokya. To his Father.*) Travelling in Central Africa is no good when you stop long in one place. A month in one camp, even in the most beautiful and interesting place, is quite enough; after that every day becomes more wearisome.... I have a lot of seeds of lobelias and 'everlastings' and other plants from 10,000 feet and upwards which I think ought to grow well in England. I wonder if you are damp enough at Flax Bourton? The lobelias are very handsome things and would be a great ornament to any garden if they could be made to flower. People in England are crying out for bulbs from here, but alas there are none, or none that I can find. Crocuses and daffodils from Ruwenzori would be worth having but I am afraid they are non-existent. There is a pretty pink and white amaryllis, but not so fine as many that are grown in hothouses in England. I undertook to surrender to the B.M. all the plants and insects that I collect, but my conscience does not forbid me to keep a few samples of seeds.

In a few months this district will probably become a part of the Congo Free State. For some unknown reason, the thirtieth parallel was made the boundary between the C.F.S. and Uganda, and when the Commission was surveying the Uganda—German East Africa boundary, they discovered that the thirtieth parallel had been put about 12 miles west of its true position. If they persist in using an imaginary line instead of the natural features of the country the boundary will run on the east instead of the west side of Ruwenzori and we shall lose the whole mountain range. The Semliki river would of course be a much better boundary, but it is doubtful whether the Belgians will see this. So far as one can understand, the natives prefer British to the Congo rule, but it is very difficult to find out the truth of anything in this country....

AUGUST 30. (*Somewhere on the road between Toro and Entebbe. To R. E. Martin.*) ...This valley swarms with elephants—they root up the trees, churn up the paths into almost impassable mud, trumpet piercingly at night and smell abominably at all times. But you very seldom see them. You may hear one crashing through the forest within a few yards of you and yet not catch a glimpse of it....The Congo forest with its damp and gloom and quiet has a very strong fascination for me, though I can quite imagine many people being horribly repelled. To feel it properly you must walk a mile or two behind your caravan, or better, sit out at sundown and hear the day things go to bed and the night things come out. I have not seen any okapis, though we passed through their best ground; and only a few pygmies—jolly little people with a keener sense of humour than most of the African natives I have met. They (the African natives in general) are abomination to me and the more civilized they are the less I like them....

SEPTEMBER 1. (*To his Father.*) As you know, we left Muhokya early in July and crossed the Congo Free State boundary (clothed this time), into a more or less unknown district with a bad reputation for mosquitoes and fever and worse things, but in most ways less black than it has been painted. There I had my first experiences of big game shooting and was glad to find that I can hold a rifle more or less straight. I am not a more bloodthirsty man than most, but I confess to a feeling of great satisfaction when I make a good shot after hiding behind a blade of grass or trying to look as much as possible like an ant heap. We made an expedition to the west side of Ruwenzori with an escort of black soldiers and a Belgian officer, and after doing more or less damage as we went along and frightening the people (I mean the escort did), the Belgian officer got ambushed by natives armed with guns and had a nasty time of it. So we were forced to retreat in haste and return to Beni. But the sad part of the story is this: I had started on an exploration to the higher parts of the valley—up that side where only two white men have ever been—and whence none have attempted to climb the peaks. I had a good chance of doing something really worth while and I was not a little hopeful. I had reached 11,000 feet with only a few hours march between me and the snow, when a messenger came up with news and I had to turn my back on the goal. I can hardly think of it now without tears. The next day—when I should have been nearing the top—the peaks were clearer than ever I had seen them—just to spite me I believe. We had an absurd time going back, a sort of running fight for three days to Beni, and when we left Beni they were preparing a punitive expedition to quell these wretched people, so the west side of Ruwenzori will not be a very healthy place for a white man for a long time to come. Since then we have been through the Congo forest to a place called Irumu, on one of the big tributaries of the Congo. I rather like this gloomy silent forest, but I can quite imagine how Stanley and the early people were half crazy with delight when they saw the sun after being for months shut in the forest. At the present moment we are rushing across Uganda by forced marches to Entebbe to enable one of our fellows to get home for the partridge shooting. It seems hardly worth while when you may shoot partridges every year of your life, but on we must go.

OCTOBER 20. (*Kasindi, Lake Edward. To his Father.*) This is the furthest outpost of the Congo State. Here, are about a hundred black soldiers, and one Belgian officer who is our host—and a very entertaining bounder. As an instance of the sort of lawlessness of the Congo, I may tell you of his performance two days ago just after we arrived. We saw a grand herd of elephants crossing the plain between here and the lake, and this Belgian asked me to come out and have a look at them, so we went, he taking a rifle 'in case of accidents'. We got close up and had grand views of the herd at less than 100 yards distance. Then he opened fire with his rifle, aiming quite indiscriminately, and after firing about a hundred rounds and wounding heaven knows how many, he by some fluke killed one. I thought I had seen enough elephant shooting to last me all my life, but when we got back to the post two wounded elephants were seen wandering about within a mile of our place, so Carruthers and I went out and shot them. They ought really never to have been shot, but now that it is done we shall take our chance and make a little money out of the tusks—perhaps five or six pounds each. The extraordinary thing is that it is possible for a State official to go out in the close season and kill elephants in a reserve where no animal, least of all an elephant, is supposed to be killed. It is typical of the whole country.

To-morrow we go off in canoes down the west side of the lake; thence three or four days to the volcanoes where we shall stop for a month or more if it repays us. Butterflies are scarce, but I have a host of beetles, some of which I hope may be good. I want to get some waterlily seeds but unfortunately they don't seem to be in this lake; perhaps I shall find them in Kivu.

NOVEMBER 19. (*Rutchuru. About* 50 *miles south of Lake Edward. To his Father.*) We ought to be a good deal further than this by now but have been delayed by rather a bad attack of fever which has lasted now for a month. We came here to recruit a fortnight ago and ought to be able to get away in a few days. It is a high-up healthy place where all sorts of European luxuries grow, such as potatoes, cabbages, lettuces and strawberries. I have a plate of strawberries every morning for breakfast; think of that in mid-Africa!...No place is without its drawbacks, and here it is lions, which, without exaggeration, swarm. They seldom roar, but grunt and growl about the place all night, and in the morning you find their footmarks within a yard or two of your door. I confess I am in terror of them, in spite of the security afforded by sentries armed with loaded rifles.

Passing through the Mfumbiro range of volcanoes, he and Carruthers canoed down Lake Kivu from north to south, about 70 miles along the western shore of Lake Tanganyika, across to the Congo River, and so down the whole course of that river to the sea.

Soon after his return to England A.F.R. wrote to J. H. Clapham and said:

I really believe that a wife is the only thing that could keep me at home, and that for how long, I wonder? Perhaps you can tell me? So in a year's time you may expect to hear either that I am married or off again to Rumtifoo. At the instance of a literary friend I am occupying myself with the production of a book (suggestions for a title, please) about Africa.* I don't suppose that it will ever get so far as being submitted

* This was eventually published under the name of *From Ruwenzori to the Congo*.

to a publisher, but the writing of it affords me a good deal of amusement, not to say pleasure. The worst part is the horrible manual labour of putting pen to paper. If I had written on paper all that I have written in my head when walking along roads, sitting in ditches, or on the tops of mountains, the second-hand bookshops would be twice as full of rubbish as they are at present....

VI

TWO NEW GUINEA EXPEDITIONS

In October 1909 A.F.R. was appointed medical officer, entomologist and botanist to the British Ornithologists' Union expedition to Dutch New Guinea. The leader of this party was W. Goodfellow, and the other members were: W. Stalker, G. C. Shortridge, Captain C. G. Rawling, and Dr Eric Marshall. One of the objects of this expedition was to reach the Snow Mountains (Nassau Range), and chief among the misfortunes encountered by the party was the selection of the wrong river, the Mimika, as a means of approach to these mountains. The great physical difficulties of this unexplored part of New Guinea, the evil climate and consequent sickness, made the work of the expedition extremely arduous. Misfortunes befell them from the beginning, and in A.F.R.'s own words: 'To write a true and complete account of the expedition would be to fill pages with repeated stories of rain, floods, sickness, and consequent inaction'. In the book he wrote on his return to England, *Pygmies and Papuans*,* he describes the expedition's important discovery of a pygmy people, known as the Tápîro, and together with an account of the appalling difficulties of travel, he tells of the bitter disappointment they all suffered in not setting foot on the Snow Mountains. He said: 'This was aggravated by the fact that we had been so long in sight of them....It was more than tantalizing to see day after day those virgin peaks so comparatively close at hand and yet as unattainable as the mountains in the moon. I looked and longed for fifteen months, and at the end, with an aching heart, I had to turn my back and leave those fields of untrodden snow as remote and mysterious as they were when we came out'.

So much was this so, that before A.F.R. left New Guinea he had definitely begun plans for returning to it. His diaries on this first New Guinea expedition were used largely in the making of

* Published by Smith, Elder and Co., 1912.

his book, and for this reason I am omitting them, as I did his African diaries, and only publishing some of the letters he wrote home at the time. I am of course giving the later diaries kept when he went to New Guinea in 1912, as his experiences then were not dealt with in any book.*

DECEMBER 19. (*Batavia. To his Parents.*) We are making for the Mimika River, on the south-west coast of New Guinea and not very far to the east of the Noord River. This is directly south, and about 80 miles as the crow flies, of the highest peak in the range yet seen—Carstensz, about 18,000 feet. We hope to get a long way up it and not find that someone has been before us. There are several other expeditions in D.N.G. now, and the one that is most likely to cut us out is conducted by one Lorentz, who has gone up the Noord River hoping to get to the Snow Mountains. They landed in September, and it was on their account that the Government would not allow us to land before January 1. Another expedition has gone in from the north coast, but they have a very long way to go. The people at the mouth of the Mimika are reported to be perfectly friendly, so I do not suppose we shall have the least difficulty with them. Our only difficulties will be with the steepness of the country and the denseness of the vegetation, both of which are certain.

JANUARY 7, 1910. (*Mimika River. To his Parents.*) We arrived off this coast on the 4th and have fixed upon a place for our camp about 10 miles up the river and have been busy landing all our gear. The process is rather a tedious one, but if the sea remains calm we hope to get it off in two or three more days. Meanwhile some of us have gone up to the camp and others are waiting here on board. There are two or three big villages, one at the mouth of the river and others higher up near our camp. The people are very friendly and not at all

* A.F.R. read a paper before the Royal Geographical Society after his return from this expedition to New Guinea, and it is published in the *Geographical Journal* of March 1914.

shy. When we first got here they came out in their big dug-out canoes and swarmed on board. They have no clothes and very few ornaments; their bows and arrows are short and simple and their clubs and axes of stone. There are always two or three canoes hanging around the ship bartering all their possessions for beads and empty tins. This morning I got a nice stone axe for a small bottle, but I shall not attempt to collect things until we come back. There are 30 or 40 miles of level country—forest and swamp—between us and the lowest hills, and then the mountains rise steeply. We can see enormous precipices which look quite sheer in places, and the whole mass—especially the highest range—looks like an immense scar or escarpment, or a great slope from the other side. Not much snow to be seen on this side for it is too steep for snow to lie. No doubt a great deal is hidden from us by intervening ridges and I expect there are very large snowfields on the northern slopes.

The flies are the drawback to this place. They lay eggs everywhere, which become maggots in an incredibly short space of time. It is now pouring with rain and the flies are maddening, so I give it up and hope that my next letter will come from a flyless camp.

JANUARY 20. (*Wakatimi Camp, Mimika River. To his Parents.*) We have begun by losing one of our party, W. Stalker, by drowning, two days after we got here. He went out to shoot alone—a thing that nobody ought to do in such a country as this—and was not missed until after sunset, by which time it was raining heavily and in a few minutes pitch dark and hopeless to move a yard through the dense jungle and swamp which surround the camp. A hundred people searched in vain for him the whole of next day, and two days later natives brought his body in. It was an experience which I hope will never again occur to me.

We get on very well with the natives here, for they are a pleasant friendly people and I have made some good friends among them by dressing the horrible sores from which many of them suffer. They are extremely primitive, practically unclothed, and their possessions are of the Stone Age. I am getting a nice lot of stone clubs and axes in exchange for bits of cloth, bottles and so on—at the average cost of about twopence apiece. The climate is not at all bad. It is excessively hot in the daytime but the evenings and early morning are fairly cool. At 4 p.m. it rains heavily, generally with thunder. Mosquitoes are rare at present. The chief drawback to the expedition is our provisions. There are no vegetables, and—what is more necessary in these places where one's appetite is poor—no variety. I regret very much now that I went off to Ireland instead of stopping to look after these things, but then it did not seem to be my business. The expedition has begun to move forward now, and for the present I am left with a Dutch officer,* waiting to send further canoe loads up the river.

We really can make no great progress until we have a steam launch which we are hoping the Dutch Government will lend us, but it cannot arrive before March, and in the meantime it will be difficult to keep supplies going. I expect all this will bore you, but to us it is vitally important. All sorts of accidents may occur, canoes upset and their contents lost; a rough sea may prevent steamer landing stores, and so on.

I have taken charge of the stores and should anything go wrong it would turn my hair grey if it were long enough, but it is cropped close with clippers, and now does not equal the length of my three weeks' beard.

MARCH 20. (*Wakatimi Camp. To his Parents.*) This is from the same address as before but since I wrote last I made an excursion with another man to the foot of the mountains and there we found a new and undiscovered race of small men. They were nine little men averaging 4 feet 3 inches in

* Lieut. H. A. Cramer, who was in charge of the military escort provided by the Dutch Government.

height: very upstanding, but shy and not inclined to be very friendly at first. However, when they saw we meant no harm they came and talked and consented to having their measurements taken. Their women and children kept out of the way but I hope later on we shall see them.

It was very jolly getting away from this low country, and the air felt deliciously cool and fresh; the water in the rivers was worth a great deal after the boiled muddy stuff we get here. I think the most pleasant thing was to feel stones under one's feet instead of the mud that we wallow in down here. I assure you that there is not a pebble within five days journey of this place.

I hope we may manage to worry through to the mountains when Goodfellow returns with coolies, but in the meantime we are wasting time and health to no purpose. Luckily we are all very well. I am bearded and rather thin but as fit as possible; only, I fear, very ill-tempered at all this waiting about. Yesterday we had a great excitement when Lorentz, the Dutch explorer, turned up after having been 150 miles east of here and having climbed one of the big snow mountains. In fact it was the first snow mountain climbed in New Guinea. The mountains we hope to reach are higher than his, and I am beginning to doubt that we shall ever get to the top of any of them. Lorentz and his party looked splendidly fit, almost as if they had returned from some health resort. They brought a lot of letters for the other people but none for me, which was a great disappointment.

APRIL 5. (*Mimika River. To J. H. Clapham.*) ... The difficulties of travelling here are greatly increased by the fact that the country produces nothing to eat. The population is scattered and not at all numerous, the people have no cultivation at all, but wander about from place to place collecting sago in the jungle or fish in the sea and rivers, and then return for a time to their villages. We have to bring in every scrap of food from outside, and so with the exception of a few pigeons (the only fresh meat we can get) we feed entirely from tins, which is neither wholesome nor appetizing....

Down here the water is half salt and mud, for we are on the edge of the mangrove swamps, and everything reeks of stinking moisture. In spite of having lost half a stone in weight I still keep very fit, but one cannot go on like this for long, and after three or four more months I shall not feel like getting up to 18,000 feet....

When we first came here these people were still in the Stone Age. They cut down trees and hollowed them out into canoes with stone axes, and broke each other's heads with stone clubs. Now we have a fine collection of stone clubs and axes, while they have a few knives and many empty tins, and go gaily decked in beads and scraps of coloured cloth. They are, of all the native peoples I have come across, the most primitive and the least interesting....They are just about worthy of the country they live in, which is from the sea to the foot of the mountains the most dreary and forbidding country I have ever seen. There is nothing beautiful in it, nothing of romance, nothing to stir one's imagination in the least, but altogether an utterly soul-destroying land. There may be something compensating in the Snow Mountains, but I have got to get there first to know that, and at present they seem to be as distant as the moon. I cannot tell you at all when I am likely to be back and I do not greatly care. Spring is the only time of the year when I would willingly be in England and I should like to transport myself there now for a month or two. My head is close-cropped like a convict's, and my beard, trimmed to a neat point, looks not amiss—and I may perhaps bring it home for you to see....

MAY 6. (*Wakatimi Camp. To his Parents.*) Common sandpipers back in the river again. Very pleasant to hear their note, but what fools to come back to this loathsome country when they might still be away up north. I really believe I would rather be in Africa with fever than in New Guinea without fever. I think I have discovered at last why this is such a hateful and depressing country: it is the trees. Rather a sad thing for me to admit, because I have more than a little love for trees, but here they are such beastly trees—rotten and scraggy—ill-grown and dead—shoving each other out of place and obscuring the view everywhere. I think when I get home I shall go straight off to a place where there are no trees, and rejoice in an open view.*

JUNE 1. (*Wakatimi Camp. To his Mother.*) I am still fretting to get to the Snow Mountains. I thought I had quitted this place a few weeks ago, but when I arrived at our camp at the head of the river I found Shortridge so ill that I have had to bring him down here and send him off to Australia to recruit. We have just heard that a Dutch expedition has started up a better river than this and more to the east. If this is true it is a most despicably mean thing to do. They would not allow us to land until January, so as to give Lorentz a chance to be the first to reach the snows, and now they try to cut us out at the last moment. So much for us.... Things go so slow here it is enough to break one's heart!

How I should like to look in to-night and hear you play music, and smell all the good scents coming through the window....

AUGUST 9. (*Mimika River. To Arthur Hill.*) ...No great discoveries—botanical or otherwise—so far, and I fear the chance of making them grows more remote, and time goes along with very little progress made. Weather worse every month. Coolies sick or dead, so can hardly move. It took us six months to get ourselves and stuff 20 miles from the coast. The snow is 50 miles farther, with, of course, the most difficult part to be done; so you see how hopeless it is. Never

* On his return to England he went to Donegal.

was such a vile country, and if we don't get to the snow I shall regret every hour spent in it. A few good photographs, natives, trees, etc., but nothing very striking. Still on the low country, so not collecting plants except orchids (mostly very dull *Dendrobiums*) for the mustard man. What I want are some startling Alpines, but I fear they will be left for another expedition. I shall probably be here until the beginning of next year, so look out for my emaciated form some time in the spring....

SEPTEMBER 1. (*Wataikwa River. To his Mother.*) I hear that Father has been making copies of some of my letters. None of my letters are worth copying, and I should prefer that Father spent his time more profitably and the letters be read to pieces if necessary. I left Tupué (which is really called Parimaú) some days ago, and am now at a camp nearer to the mountains, but still I fear a very long way off. On the journey here I was held up by a flood at one of our camping places on an island in the river. For three days I was in fear that we should be washed away, but our only loss was one shoe. It might have been everything. Incidentally the coolies with me threatened to desert and I had to coax and promise them all manner of things to get them to come on here. For the moment we can do nothing but accumulate stores against the time when we can push on further. It is not wildly exciting, but I have two or three books to read....It is fearfully disappointing to have made so little progress in these eight long months. So much might have been avoided by making adequate preparations and by finding out beforehand the little that was to be known concerning this country. We have quite unsuitable coolies and if the next lot are no better we may as well pack up our things and go home at once, with our tails between our legs and a smile on the faces of the Dutchmen. It is a thoroughly beastly country and if I don't get a foot on to that snow I shall consider it a year of my life wasted.

I spend a good deal of my spare time dreaming plans for my projected Utakwa expedition. I read whenever I can, and an odd thing happened this afternoon. I was cursing a cicala that kept up a horrid din in a tree close by, when in the book before me I read the lines: 'creaked like the implacable cicala's cry' (*Ring and the Book*).

DECEMBER 28. (*To his Parents.*) Made a trip down to Merauke, a Dutch post near the border of British New Guinea. On the way we visited two other rivers where there are Dutch exploring expeditions, and one of them, the Utakwa, is undoubtedly the river we should have come to, as it apparently comes straight from the Snow Mountains and is more navigable than the one we are on. A steam launch cannot get over the bar at the mouth of this river at low tide, so you can imagine how hopeless it is.

Merauke seemed quite a metropolis to me after a year out here and I revelled in the unwonted luxuries of china cups and dinner napkins. I find the Dutchmen exceedingly kind and hospitable, and in a way much more like English people than any other foreign people I have met; their manners are a great deal better than our own.

There are now three of us left. I get on all right now, and no doubt we shall rub along together until the end of the expedition. I like them all a good deal better than I did at first. You cannot think how one longs to see some fresh people and to hear some new ideas on things in general.

A.F.R. arrived home in England early in June 1911, and a few weeks afterwards he writes to his mother from Donegal:

Here, in Donegal, one is out of doors most of the day—blissfully happy, and with a good appetite at the end of it; nothing to show me that the country really is going to the dogs! Let me tell you that it is almost worth while to come here for the mutton alone: it is excellent beyond words and not only by reason of the hunger that seasons it. Another thing that strikes me especially here is the beauty of the children. Every day as I go off to fish I meet little parties of them on their way to school, and in every group there are sure to be one or two noticeably pretty children. Their beauty is in no way connected with the mutton, for their parents tell me that the children never have eaten meat in their lives. Bare feet, a shawl and no hat, is a very becoming costume which might well be copied at home; but I suppose we are too proud for that. Really, the Irish do appeal to one's sympathy—if not always to one's reason.

In February 1912 A.F.R. was occupied in making preparations for a second expedition to Dutch New Guinea. He was to start in May and expected to be away not more than a year.

MAY 6. (*London. To M.M.*) ...I am horribly disappointed, but I had a sort of idea that somehow or another you would not come to tea with me on Thursday. I don't mean to say that I thought you would try and get out of it, but that something would prevent it. You haven't missed much in the way of an entertainment I am afraid, for nobody has ever been to tea with me except an occasional man about once a month. Never a woman, or lady or girl (what a pity there is no word corresponding to man) has ever been here, and you would have been a new and very delightful experience to {which / whom} I was greatly looking forward. You will think me an awful idiot, but I know so little of people—apart from men—that I never dreamt that it was even possible to ask you to come to tea in my den, and now I am consumed with regrets when I think that you might have come here (or rather that I might have invited you, which is not quite the same thing) several times in these months that I have been living here; but then I did not know that you lived in London and you were barely, if at all, conscious of my existence. What a rotten life it is, and how confoundedly short, as you will find when you come to my years of discretion! I seem to spend about half my time in saying good-bye to people whom I never see again. I hope that will not be the case with you, but if you will persist in going to visit the Kurds you stand as good a chance of coming to a nasty end as I do in New Guinea; so don't go there, please. You don't say when you are coming back from the country, but if it is before the 16th, when my books and other gear will be removed from here, you will find me ready and anxious to make tea for you on any afternoon that you can come. If that is not possible, it is good-bye for a year, and I am glad to have your good wishes for my adventure. I am still so ignorant of you that I do not know whether this is an impertinence, but I will take the risk: if ever you are bored with civilized life and you want to talk to a savage without restraint, a letter to me will be greedily welcomed. In the meantime all happiness be with you....

P.S. I see you addressed me first at the Savage Club, thinking that that was the proper place for such a wild beast as I am. I am a dreadful savage but quite harmless.

And so A.F.R. again left England for the mountains of New Guinea. This time he was joined by C. Boden Kloss, Curator of the Museum of Kuala Lumpur, and they chose the Utakwa River as the best means of approach to the mountains. In the following letters and diaries, A.F.R. tells how after months of hard travelling, partly by canoe and partly over land, they succeeded in reaching the glaciers of Mount Carstensz, attaining a height of 14,866 feet. Here, less than 500 feet below the summit ridge of the mountain, he and his companion were forced to turn back and abandon all hope of gaining the top and seeing over into the unknown country beyond.

JUNE 23. (*Singapore. To his Parents.*) I am busy making arrangements with shipping companies, dealers in rice and so on, and am trying to find a suitable man to look after the motor-boat and act as storekeeper. In this I have not yet been successful, but I have got a number of people on the look out for him and he has got to be found somewhere. I have altered my scheme in one respect and have arranged to get half of my Dayaks (forty men) from Sarawak; the other forty I shall get myself in Dutch Borneo. I am already a bit later in leaving here than I intended to be, so I think it is most likely that we shall not finally leave Java until almost the end of August—or perhaps even the beginning of September—it doesn't make much difference.

JUNE 26. Have seen all sorts of unsuitable people with a view to getting hold of a good man to look after the motor-boat and act as storekeeper. Have finally got hold of an Eurasian, George Siddons, who seems to be the right kind of man: twenty-seven, unmarried, been through motor works for five years and has run a motor-boat for one year: keen on shooting and not afraid of jungle. Have engaged him at the equivalent of £17. 10*s*. 0*d*. per month and hope he will turn out all right.

Went down in the evening to see the motor-boat, and found that the P. and O. had smashed the gunwale in getting her out of the hold; also found that an inside part of the oil pump was smashed in two. If I had not found this out till we got to New Guinea the boat would have been useless, so it was as well I stopped on here to see her before going on to Java. Mended the pump and had a satisfactory trial. Engine gets fearfully hot but on the whole my *Ibis* goes well; Siddons seems to take to the job all right.

JUNE 30. Arrived Batavia and spent the next few days interviewing officials. Am getting all my stores from the Government, but they insist on sending a huge escort with me, which is quite unnecessary and a great nuisance. The officer they have put in charge is rather a lumpish person and not nearly so nice as the man who was with us on the first expedition (Cramer). Intend going to Borneo to 'catch' Dayaks.

JULY 9. (*Pontianak, Borneo.*) Anchored outside the bar of the river by which you get to the Kapuas river. Complicated network of waterways here. Pontianak actually is only a few miles up a river from the sea, but we have to go by cross channels about 80 miles to reach it. At midnight, being high tide, we started to cross the bar into the river but went hard aground and could not get off again. We shall have to try again this evening when tide is high. There is another steamer, about the same size as ours, which has been stuck on this bank for four days and is now waiting for spring tide.

Got off the sandbank yesterday and steamed up the Borkoe river. Nipa and mangrove swamps, then occasional coconuts; once or twice rubber. Into the Kapuas river—very wide—and then into the Kapuas Kechil river, and so to Pontianak. Went ashore and interviewed the Resident, Harbour Master, and others. Walked about the town and saw mostly Chinamen who are not at all attractive now that they have all cut off their pigtails. They try to look like Europeans and give themselves most absurd airs. Pontianak is built on, or rather in, the river. Canals run everywhere between the houses, and the water rises and falls, leaving a most malodorous mud at low tide. Dirty place, but seems to be prosperous. Oil and timber factories, and many Chinese steamers of small size. I put up at a grubby hotel kept by quite decent folk. Under the eaves, just outside my window, nests a colony of common white-rumped swifts. They are always very busy in the evening and keep up a constant twittering long after they have gone to roost.

There are no horses here and no wheeled vehicles of any sort, except one or two rickshaws kept by rich Chinamen: everybody goes everywhere by water in sampans.

JULY 12. Left Pontianak in a Government house-boat, towed by a small steamboat drawing about 5 feet of water. Pleasant way of travelling but not speedy. A few hours from Pontianak we stopped to take up firewood and I saw quantities of monkeys: also noticed the hornbills here are different from those in New Guinea. This Kapuas river is immensely wide, with a strong current. The jungle is ugly and very dense—might be New Guinea; occasional Malay villages. Stopped at a large rubber estate called Sunidekkan, and then went on to Sanggan, a large Malay village. Busy trade here in 'tingkawang'—a native nut from which much oil is taken and shipped to Europe, where it is used to manufacture soap, butter, etc. At Sintang the Assistant Resident advised me to try and get Batang Lupar Dayaks and not the Kajans. Everybody tells me a different story about Dayaks and it is hard to know what to do.

JULY 12. (*Pontianak. To M.M.*) ...You said you would send me a copy of your *Cornhill* article, but it never arrived, and I should never have seen it had not the (usually) faithless Reginald Smith sent me a copy which I found at Batavia ten days ago. So I will magnanimously forgive you for having forgotten me and your promise, and I will heap coals of fire upon your head by congratulating you on having written such an entirely admirable article. It is curious how much better you seem to know a person when you have read anything he (or she) has written. I have found out all sorts of things about you that I did not know before, but I am sure there were lots of other things in that other part of your diary which you suppressed. I should like to read that too, but I don't suppose you will ever let me. One thing is quite certain, and that is that your brother had an ideal travelling companion, and I hope he was aware of that fact. I wish we were not quite so horribly conventionally civilized, and that you could transport yourself on a wishing carpet to these parts and accompany me on my wanderings. I know that I should see twice as much and get double the amount of pleasure out of it that I do alone, but perhaps you would not get quite the fun out of Malays and Dayaks and Papuans and the other queer people that I have to deal with that you do out of Kurds and Georgians and such folk. Anyhow, I am afraid the days of wishing carpets are past and I shall continue to wander alone.... This is a very queer sort of place, built apparently on a river, with the houses surrounded each by canals in which the river rises and falls, leaving at low tide a horrible mud which offends my too sensitive sense of smell. Most of the inhabitants are Chinese, and together with their pigtails they have abolished their glorious yellow flag with the dragon, and have substituted for it a hideous monstrosity of five stripes of quite incompatible colours. Talking about stripes reminds me of the American lady I met in Singapore. I had been trying to explain to her that in New Guinea there was nothing to eat, and that all the food for me and my eighty men would have to be carried in tins. The only comment she made was, 'Oh, Mr Wollaston, what tons and tons of ice you will have to take with you!' I am afraid I must be an awful idiot at explaining things.

By the time this gets to you, you will no doubt be off on some journey to the other side of somewhere, and I shall imagine you sailing down the Euphrates or getting captured by brigands. I hope not....

JULY 17. To Smitau, and on in the evening to Selimbouw, a very interesting place—real Malay, with wooden houses on piles over the river. These last three days I have crossed the equator about sixteen times. Took a sampan and went to look for monkeys. Very soon came across two of the proboscis monkey. Shot one and picked it up with some difficulty; a most hideous beast—very large with remarkably small head. Not fully grown so the nose was still turning upwards: only the old ones have the long drooping nose. Saw several other monkeys, long-tailed and short-tailed. Lots of crocodiles here, and I have been shown a very ingenious hook for catching them. A monkey is the bait, and when it is swallowed they pull on the rope—to which is fastened a long rattan—and the sharp blade rips up the throat or stomach of the crocodile.

Terns (species?) flying about in pairs everywhere.... The rivers here are really channels through the belts of mangroves which divide one big lake into many. We anchored at the top of a mangrove and went in a sampan through a winding channel cut through the trees until we came to solid ground. Went to a Dayak campong. In a Dayak campong you do not ask how many houses there are, but how many doors. It consists of one large house on piles reached by a steep ladder in the middle. One side of the building is a long half-open verandah, and the other half is divided into sleeping rooms with doors opening on to the verandah. I saw bunches of human skulls, old and blackened, hanging up from the roof. Unfortunately most of the Dayaks were away, and on speaking to some Dutch missionaries I was told that these Batang Lupar Dayaks are of no use whatever, and that I should do much better to get Dayaks from the Embaloeb district to the east of this place. These missionaries have a school of about twenty Dayak boys—very jolly little beggars—but they all revert to the primitive Dayak state when they have passed school age. They are very different from Malay children, for they are always doing something, chopping wood or playing games. They have a game rather like an elaborated 'touch', and an excellent game for practising spear throwing. Two or three boys stand at either end of an open space about the size of a lawn tennis court. They then throw a spear in such a way that twisted rattans attached to it bounce rapidly along the ground in the shape of a hoop. The boys at the other end of the court then have to throw their spears through this hoop as it goes along. It is most skilful. They wear their hair long behind and cut straight across the forehead; it is black and straight. They have faces something between a Chinaman and a North American Indian—very bright and intelligent.

A fellow who is with me has just been trying a new method of shooting, as practised in Celebes. Two men, one with a gun and the other with a lamp with a very strong reflector, walk carefully through cultivated ground at the edge of the woods from which deer come out at night. The holder of the lamp moves it carefully from side to side, quartering the ground until he sees the light reflected in the eyes of the animal which show up like two little bright lamps. Whilst the deer's eyes stare fixedly at the lamp, you slowly approach, prepared to shoot. Unfortunately the man holding our lamp was not clever in keeping it steady and the animal we found sprang away. It was a great pity, for I should like to have seen it successfully done, and I am told that you can walk up to within 10 yards of deer and shoot them so. Of course you must aim between the eyes, as there is nothing to indicate the position of the animal's body. Pigs' eyes look red in the reflected light, and they will not wait like deer.

JULY 19. Set off towards the Embaloeb river to look for Dayaks. We tramped through jungle on to a steep ridge of hills, and then went across some pretty mountain streams, on through hot scrub and out on to a blazing hillside; then down again to primitive forest. In one part we progressed like a troop of monkeys along felled trees, for these Dayaks can fell trees in such a way as to make a continuous bridge for more than 2 miles through the forest. It is all very well for them with their bare feet, but I find it very difficult in boots to keep on doing balancing tricks on slippery tree-trunks, in some places 12 or 15 feet above the ground. Eventually we got to a large Dayak campong called Oekit Oekit. This campong was about 80 yards long, raised on very high piles perhaps 40 feet from the ground to the top of the roof. We mounted by a ladder polished brightly by the feet of generations of men, and there was a deep hollow worn in the back of each step where native toes rub against it. It was a great disappointment to find that most of the men were away, and I think the Dutch Contrôleur who came along with us was a pretty hopeless fool, for he did not know anything at all about the Dayaks of his district. We met a young Dayak, and after talking to him a bit I asked him if he was married. No. Was it because he had no money? No; he had money, but first he must get some heads. I was rather astonished, as he was comparatively civilized and also a protégé of the Dutch missionaries!

JULY 20–22. Back to Smitau, and up the river to a place where I am going to wait until a man called Daim returns from visiting other Dayak campongs. This man Daim is a half-Dayak recommended to me by the Assistant Resident at Sintang. He was a mandoer of the Dayaks on the Dutch expedition to Utakwa and Island river, but had trouble with his commander. He is coming with me at the huge salary of fifty guilders a month, on the understanding that he has nothing to do with the military part of the show; he is well spoken of by everybody and I believe he is a good man.

Had some shots at several crocodiles, the long-nosed fish-eating kind as well as the other.

Took up quarters in a rest-house at Boenort with the prospect of two or three weeks with nothing to do. There is an epidemic of cholera round here and during the last six months they tell me 300 people have died in the district, so we must be very careful about water.

Extraordinary number of fish here, and I have seen some very big ones jumping clean out of the water in pursuit of smaller ones. Wish I had a rod and spinning bait. These Malay natives are very clever in the use of a casting net, weighted with stones or a ring of chain round the lower edge. Two men go in a canoe, one paddling in the stern, and one standing up in the bow with the net folded on his right arm and a small piece of the lower edge held in the left. They very quietly approach the place where fish are seen, and then the netter makes a wide sweeping swing without in the least disturbing the balance of the canoe, and casts the net in a perfect circle some 4 or 5 yards wide. Nearly every time he gets something, and I have seen a man get as many as twenty fish of about a third of a pound each at a single cast.

JULY 25. (*Boenort.*) Very tedious here with nothing to do; impossible to walk more than 50 yards from the house on account of the swamp. It is so hot that I do not go off the verandah till five o'clock, when I take a canoe and paddle about the river until sunset. When I go out in a sampan the people call it 'main sampan', i.e. 'play sampan' or 'playing at sampans'! They are rather astonished at my knowing at all how to do it, and they think me rather a fool and inferior person to paddle myself.

Learnt rather a nice Malay expression this evening when I was going out. The man in the next house asked me where I was going. I waved my hand vaguely about the river, and he said 'Tuan goes to take the air' (Makan angin)—literally, 'to eat the wind'.

JULY 29. (*Boenort.*) The perfectly appalling smell here is now accounted for by a dead and horribly inflated crocodile floating down the river. The river has fallen very much, about 8 to 12 feet since I came here, and the dirty water eddies round and round in a slow pool—none too healthy.

This afternoon the ex-Rajah of the place, quite a decent person (more Arab than Malay), called on me and asked if I would come and see a sick man in the village. I went across and found a poor wretch far gone in dysentery. For four days they had fed him on rice, and when I told them that that was the very worst thing for him they said that if a Malay could not eat rice he would die. I am afraid he will either way.

It is a bit tedious talking Malay, in which language I am not very proficient, and I have now exhausted most of the subjects my Malay will run to, such as—the fish in the river

—the Dutch system of taxation—the Chinese Republic, and so on.

There are signs of petroleum here in the river and gas is always bubbling up through the water. Last night we shoved a bamboo down into the sand at the edge of the river and lighted the gas that came up; it burnt all night with a flame like that of a spirit lamp.

A little rain this evening—only the second time any has fallen since I came here. Went across the village to visit my third 'cholera' patient, happily the first two are getting on all right and the people seem to be quite pleased with me. Walked through the Chinese part of the place and found it much more orderly and well-built than the haphazard houses of the Malays. The Chinese have their own school but the Malays have none—although this is a Malay country. I can't quite see what the Dutch do for the people in return for the taxes which they pay.

AUGUST 1. Went across the village to see the old mother of the ex-Rajah. She has been taken very sick—hopeless, and I can do nothing for her. There were about forty people in the house waiting for her death. She died after only ten hours illness, and I cannot think but that this epidemic is cholera.

AUGUST 2. Great day. A small Government steamer has just come up bringing a fine mail for me; also my man Daim has turned up with news that he has got forty-eight Dayaks. Thirty-six of them from Batang Lupar, and twelve from another district—all anxious to come to New Guinea. They live in the neighbourhood of Bakoel, so I shall have to go back by Smitau, Landjak and Bakoel to fetch them.

AUGUST 4. (*Boenort.*) The people hereabouts constantly send little houses floating down on the river—small models of the ordinary house, about 4 to 6 feet long—decorated with flags. Inside they put offerings of rice and money, and so far as I can understand it is all with the object of averting the cholera which is beginning to cause a great scare throughout the country.

(*Sintang.*) The evenings are particularly lovely here, for the river is arranged exactly to suit the sunset. The six Europeans here seem to be of no use to each other, as each lives in his own house and rarely appears outside. They do not dream of playing any kind of game, or of going out to shoot or fish, except the doctor, whom they all consider a bit mad.

A few days later he goes to Bakoel and Landjak and visits a Dayak campong. He persuades the people to show him all their things: blow-pipes, poisoned darts, spears and drums. He finds them industrious folk, all very friendly but not over-clean. 'They are a great deal stronger and more energetic than the Malays who affect to despise them. They still hunt heads, but it is many years since they took a white man's, so I have no fear of my own.'

AUGUST 15. I was invited to go to a Dayak house and see the people dance; so, after dark, I went up river in a canoe and mounted a large house crowded with people and lit by torches. I was provided with a chair in the most conspicuous place and then was pressed to drink various kinds of drink. It was explained to me that they were of various kinds, but they appeared to me to be all the same and all equally nasty—like ginger beer that has got stale and bitter. Most of the men sat with their backs to the wall close together while the women moved about as unobtrusively as possible, carrying drink in bowls and cups of different shapes and sizes; they sat down in front of the men while they drank and the good manners of the men were quite noticeable: there was no hustling for drinks—every man took it in his turn. They certainly consumed a considerable quantity, but as long as I was there—about three hours—the only man who got at all uproarious was one of the 'pradjurets' (native Malay policemen) of whom there are three stationed here to keep the

Dayaks in order. The entertainment began with a concert of drums, wooden and metal, evidently an imitation of the ordinary Malay 'gamelang'; then a chorus of eight women came and stood in front of me and sang in my honour a song with innumerable verses, each verse ending in a kind of little cheer. This was interpreted into Malay for my understanding, and as I understood it they addressed me in very complimentary terms; said how glad they were to see me, how they hoped I would stay with them, how they would all like to go to New Guinea with me, how they would like me to be their Rajah and so on, and in one verse I think they supposed that 'Tuan Allah' must be something like me. After they had finished with me they went to every man in turn and sang a verse for him.

When this was over, a plank about 10 feet long and 5 inches wide was put down on the ground and the performers began to dance. I have never seen anything like it, and I have never seen anything so graceful. Every man's dance was a solo performance, and every man was almost perfectly made, so that in the half-light of the torches, with deep shadows behind, these dancing men looked very beautiful. They danced to drums, or to the music of a reed instrument (with quite a pleasant tone), and they more or less kept time with the music. I cannot describe their dancing, but it was rather a succession of different bodily attitudes than dancing in the ordinary sense. They could spring from one end of the board to the other in the most wonderful bird-like bounds, and they could move slowly along it, varying at every forward movement the curves of their limbs and body. I shall never forget the beauty of their arm movements. Later in the evening they became excited and did warlike dances, wearing feathered hats and curious cloaks of monkey fur, and with a 'parang' they imitated scenes of fighting and killing an enemy, finally cutting off his head. One old fellow—a splendid dancer—did a comic performance in which he pretended to cut down trees, make canoes, plant and mill rice and so on, which sent all the children into fits of laughter. Most of the dancers were quite young men and it was amusing to see their bashfulness when they were called upon to dance. They said they could not, and had to be dragged from their places to the dancing plank, but when they got there they performed very well indeed. Altogether it was an extraordinarily interesting evening, and I find it hard to reconcile their artistic dancing and courteous manners with their undoubted ferocity. Bunches of human heads were hanging from the roof above us and they are always ready to add to their collection in their tribal fights.

AUGUST 16. Last night's entertainment has ended in a tragedy. This morning I was asked to go over to see an old man, very sick in the stomach. I found him already half dead and quite beyond any help, but it was difficult to make them understand that. They said he was quite well yesterday, drank wine in the evening, but was taken violently ill with pains in the stomach during the night. He died this afternoon. They think it is cholera, but of course it is not.

AUGUST 17. I now have forty Dayaks. They are very jolly fellows and they look strong enough for anything. I walked back to Landjak with the whole crowd, including women and children and friends and others who want to be allowed to come too. I believe there would be no difficulty in getting a hundred men in this district and so I now wish I had not been persuaded to get men from Sarawak. On the way through the forest I saw to my huge delight a Gibbon (*Hylobates*) swinging along the branches of the trees....

I have paid the Dayaks a month's advance, 22½ guilders (a monstrous amount for people of this kind), and when they got the money they went straight to the Chinese store (there are two in this place which consists of only three houses), and they bought themselves the most outrageous clothes in which they look perfectly frightful. It was quite funny to see them come out of the store, grinning with pride and being chaffed by all the others. Most of them never owned a coat or a pair of trousers in their lives, and it is a pity they have them now.

AUGUST 18. Left Landjak and had a tremendous business getting our boats through the shallow water of the lake, which has now almost disappeared. The Dayaks paddle well and enjoy racing one boat with the other.

When I got to Sintang I received a telegram saying that the Sarawak Government refuses coolies, and that only eight 'Dayaks of sorts' have been engaged. I really cannot understand this, for I had a letter from the Rajah there authorizing me to recruit Dayaks, and I have been counting on forty men from those parts. It is a frightful blow to find myself now with fifty men (of whom eight are an unknown quantity), instead of eighty. Moreover it is impossible to go back and get more from Bakoel, as it would mean delaying everything and it would be a fearful expense; I am at my wit's end.

AUGUST 23. Came to Sunggai Dekan, the English rubber estate, where they have a number of Batang Lupar Dayaks engaged in clearing the jungle. When I was here last, these men told me that they wanted to come with me to New Guinea, so I am trying to persuade the manager to let me carry them off. He utterly refused at first, and has only given in because I have undertaken to pay the expenses and transport to his place of fifty fresh Dayaks—a very costly business for me and a very good bargain for him. This manager has thirty-six men, and they all want to come with me but many of them are quite unsuitable. They say they will not come away to-night for they must dream on it first, and if their dreams are good they will start in the morning.

AUGUST 24. This morning all the Dayaks had dreamed well except one man, a particularly fine fellow. It is a pity, but he dare not come. I lined them all up and picked out twenty-four, of whom sixteen are good men; the others may be of some use so I will take them as well. One is quite a boy and very weedy, but I had to take him as his brother would not leave him alone. I now have seventy-four men of sorts, fifty-eight being good men. I wish I had stuck to my original plan of getting all the men myself in Dutch Borneo, and I shall be careful not to trust any other person's judgment in future. So far, this is the only thing I have deputed to somebody else in the preparation for the expedition, and it is the only thing that has gone wrong.

AUGUST 25. Back at Pontianak where I hear from Batavia that the Dutch Government will take us all the way from Java to New Guinea in their steamer, which is a very good business as it means a saving of about £200, and it avoids all the risky business of having Dayaks on a packet steamer with other passengers, and all sorts of possible trouble. Meanwhile, my Dayaks are lodged in the prison here at thirty-three cents a day, and they are still buying most extraordinary clothes, nasty scents and cheap soaps. One man has bought himself a huge clock for four guilders. They are very jolly and I am getting quite fond of them.

AUGUST 26. Left Pontianak in a beastly boat with very Dutch passengers. The Dayaks are tremendously excited with all the arrangements of a big steamer. What they like most of all on this ship is the steam whistle, which always makes them laugh.

On September 12 the expedition sailed from Ambon. It consisted of C. Boden Kloss, Siddons, A.F.R., 2 mandoers, 2 boys, 66 Dayaks, 13 Sarawak and 7 Malays (Butonese), making altogether 93 people. The Dutch escort consisted of about 132 men, which to A.F.R.'s mind was 'far too many and damned too much baggage'.

SEPTEMBER 16. Anchored about 8 miles south of the Utakwa. Sent steam launch to look for channel into the river. It reported 20 feet of water at bar, so we anchored up and got off only to find 14 feet and the tide falling rapidly. Had to go back again for the rest of the day and thought it was a great pity to waste time like this, but was somewhat rewarded by a fine view of the snows of Carstensz at sunset. I think I only saw them four times altogether in the afternoon before.

SEPTEMBER 18. Came into the Utakwa without any difficulty this morning, and it was lucky I had been here before as the captain was very nervous and did not like coming into the river. Sent the steam launch and motor-boat towing two boats full of Dayaks to clear a camping place. The river just like the Mimika but fine and big. We cleared a good space of ground, and though it had been occupied less than two years ago by Van der Bie's expedition, an extraordinary amount of growth had taken place meanwhile. The Dayaks delight to bathe again after the filth on board the ship. In coming away both the launch and motor-boat got stuck on to a sandbank and we found a lot of sand and grit had been driven through the water cooler of the motor-boat and had smashed the spindle of the pump. Siddons made a temporary job of it, and as it was then quite dark we had to go very cautiously down the river and I did not at all enjoy the business of steering. Round a bend close to the bank I saw a huge tree slowly tilt over and fall with an enormous splash into the water—about 100 yards ahead of us. They have a nasty habit, these great trees, of suddenly toppling over when there is not a breath of wind. Though we often heard them fall at night on the Mimika I never saw one actually crash before. As we went along down the river I saw a tree brilliantly lit up by a swarm of fire-flies, so bright that the tree was clearly reflected in the water; it had a peculiarly fairy-like appearance which I should have appreciated more under other conditions. We were lucky to get back to the ship at 10.30 p.m.

SEPTEMBER 19. Wasted a lot of time mending the motor this morning, but eventually got off towing two boats with Dayaks and baggage. Arrived at our camping place, and in a few hours everybody under canvas and busy cooking food. Luckily not a drop of rain all day—a thing simply priceless.

Busy making camp and clearing ground. Find the Dayaks don't like the tents we have got for them so they have made houses of their own kind roofed with pandanus. I gave them two days to do it in and they have made a very good job of it. They are really very jolly fellows and I like them immensely—and I think they like me which is all to the good. They work very well, but with frequent interruptions to bathe or smoke. I don't urge them on too much because that led to trouble with the last lot of Batang Lupar men who were employed on an expedition. They consume a perfectly appalling quantity of food, one kilo of rice a day besides fish, and a large amount of odds and ends picked up in the jungle. We have had two or three fine moonlight nights and after food they like to dance or have jumping competitions, doing tricks of bending backwards and so on; they wrestle splendidly, and every fall is greeted with shouts of laughter. They enjoy themselves immensely and I am never tired of watching them.

SEPTEMBER 28. The captain of the *Valk*, which brought us here, has given us all a very splendid feast, and as the ship has now discharged all our stuff he is now sailing away, leaving the whole expedition at base camp.

A few days ago we were visited by some Papuans, but they were shy, and at first would not come out of their canoes; not at all like the Mimika people who crowded into camp the first day. I think I recognized one of the men who had been to see us at Mimika, but he would not come and talk to me. In the evening some Dayaks came and asked me if they might take some of their heads.

These Dayaks are almost as good as the Papuans at managing canoes, but the canoes we have are a bit too small to use for carrying baggage, so we must set to and make new ones or else buy big Papuan canoes. Unfortunately there is a dearth of big trees just here, and we have found only three good enough for canoes and they are at some distance from camp.

Weather fine on the whole and river low.

OCTOBER 2. To-day we induced some Papuans to come into camp. I gave each of them a piece of red cloth, so they were soon quite at home. They seem not to understand a great many Mimika words.

Heavy rain last night and the river in flood, so we went up in our motor-boat. Got entangled in a lot of wood and then stuck on to a sunken log; had a good deal of trouble in getting off but luckily no damage done. We killed a fine cassowary and skinned it before night.

OCTOBER 8. (*Utakwa River.*) The Dayaks are busy making canoes and in three or four days we ought to have five finished. Like the Papuans they cut a tree down, hollow it, then very cleverly open it out by fire. The process is quite simple. When the log has been hollowed out, it is turned upside down (supported on trestles about 2 feet from the ground) and a long fire of dry wood is laid underneath to char the inside. In a few hours the log is turned over, and with a smaller fire, directed more to the sides than to the middle, the outside is charred. At the same time, with an arrangement of hooks of wood, pieces of rattan and weights, the log is slowly opened out into the form of a canoe; i.e. from a shape somewhat like a horseshoe it becomes a slight curve. The inside is then well scraped, thwarts are fixed to the brackets and long planks are nailed on to the gunwale of the boat, each one made from a single tree; these increase the capacity of the boat at least two times. The Dayaks then paint the outside of the boards, fill up the open bow and stern with wide triangular boards, and carve an extraordinary beast for a figurehead which makes the boat complete. A boat can be made like this in ten to twelve days, and there are generally thirteen or fourteen men working at each boat. They seem to like the work immensely and get on with it steadily. They have given names to their canoes, such as 'Sapoerantan' (The Sweeper of the Reaches), 'Bordjang baleh' (The Return of the Bachelor), 'Bordjang Kilat' (The Flash of Lightning), 'Makan Tandjong' (The Eater of the Bends, i.e. of the river), and so on. I like these Dayaks very much and I wish I had more of them. Of the sixty-six I have, only about fifty are fit for very hard work, and it is galling to know that if I had followed my own plan I might have had eighty of the best.

At this last full moon we had two or three beautiful nights, and this inspired dancing and music and wrestling, with all manner of queer gymnastic tricks. They are all as happy as children. Perhaps they will not be so cheerful when we get into the mountains, but they will not have to stop there for any length of time, for as soon as they have carried their loads up they will come back here. My plan is to go up the river as far as possible, make a big store of food, and then go for five or six days' march to where we shall make another store of food, and so on again; each store becoming smaller and of course more difficult to make as we get higher into the mountains and further from our base store. The weather is kind: sunshine in the afternoon and fine evenings—not at all typical New Guinea weather.

The natives I have seen here—there are very few—come, I think, from far off, and do not speak quite the same language as the Mimika people, neither do they appear to be so intelligent. I hope we shall find people living in the mountains.

If so they ought to be interesting, though I do not think we shall find the pygmies up there as they do not come so far east.

OCTOBER 14. Van de Water* and I, with Dayaks and soldiers, went up the river to-day in canoes, but after about 3 miles we found the water so low that we were more often out of the boats than in them, hauling ourselves over logs and stones, cutting and chopping and incidentally having a continuous bath. The Dayaks seemed to enjoy themselves immensely, although there was a blazing sun and not a breath of wind. We searched many hours for a camping place and hoped to find signs of Van der Bie's camp, but it must have been quite washed away. Saw a few natives in scattered groups, but no women or children and no permanent houses. The river here varies from 40 to 200 yards wide, with here and there islands; water clear but very warm.

OCTOBER 15. Off early, and struggled on much the same as yesterday. Great pity the weather is so dry, as the mountains are entirely hidden in mist and we have not seen a glimpse of them. I love to watch these Dayaks enjoying themselves, in and out of the water—chasing fish whenever they can. Once a large water tortoise was seen close to my canoe, and in a minute all the Dayaks were overboard with spears and paddles, swimming under water and giving wild chase to the creature, which eventually escaped. I think they would have stayed under water all day—with an occasional foot or blowing face coming to the surface—if I had not literally dragged them on. After eight hours' paddling we found the place where Van der Bie had had his camp, overgrown with dense and tough bush. We set ours on a shingle bank. The river is strong and stony here, and coming up my trousers prevented me from wading through the rapids, so I cut them into 'shorts'; in consequence I am now horribly burnt on my legs, which are a mass of blisters. But it was

* The officer in charge of the Dutch escort.

almost worth while, for half the time I enjoyed myself wet up to the neck—which is a thing I haven't done for ages.

Cleared the ground and returned to our base camp. Saw a splendid cassowary strolling along the sand in the most casual way and taking no notice of us; of course I had no gun.

OCTOBER 22. (*Base Camp.*) Kloss has gone up to the top camp with five canoes.

We were visited to-day by a crowd of men from the Neweripa river; shy at first but they soon came on shore and traded axes and so on—nothing of interest. They pointed out two men in one of their canoes, both wearing bone ornaments in the septum nasi and both obviously different in appearance from the others; the people called them Tápîro, which excited me a good deal, but they are not the same as our Mimika Tápîro. It is a great pity that there is no village on this river.

This afternoon some Dayaks, who are making a canoe near camp, rushed to me in tremendous excitement saying that some Papuans had stolen two of their parangs. In an instant they were all off with parangs and spears to chase the Papuans, and just as they were getting into a canoe I ordered the whole lot out of the boat, which caused great fury against me. They shouted and cursed, and one man in his rage smashed the figurehead of the boat (hideous monster carved from wood). When they were a little quieter I told them I was coming with them; this made them happy again. Thirty-two Dayaks, a Dutch sergeant with a gun and I, all crowded into a canoe and went hell-for-leather down the river for about 4 miles, but the Papuans were easily ahead of us so I persuaded the Dayaks to turn back. They were so furious that they would certainly have done some damage to the Papuans if we had come up with them; indeed they are asking me now if they may not take the head of any Papuan who comes along. Great pity that this should have occurred, for the natives will now be shy of visiting us.

OCTOBER 24. (*Base Camp.*) I am making a great clearing across the river so as to be able to get a view of the Snow Mountains from this camp. One of the convicts who is helping in the work chose to-day to swim across the river instead of going in a boat, but very soon after he had started he failed and sank. Someone rushed in to tell me, and I sent off Dayaks in a canoe in less than no time, but the man was not seen again—only a stream of bubbles in the deep eddy just at the corner of the camp.

OCTOBER 26. This early morning I had a good view of the Snow Mountains, and when a few more trees have been cut down it will be fine indeed. I went up the river again, and at one difficult rapid one of the canoes ran into a snag, smashed its side, filled with water and turned over. The baggage which began to float away was quickly recovered, but several tins of rice, and more important still, chocolate, milk, soap tablets and other valuable things, sank to the bottom and could not be recovered; fearful loss. We put the recovered baggage in the other prows, which were already well loaded, so it meant a very heavy time getting up to camp. Rewarded by a fine evening and no mosquitoes.

OCTOBER 27. When I got up this morning to our top camp I heard a dreadful piece of news. It was that a few days ago one of a party of Papuans visiting the camp had stolen a pick, and was making off down the river when our sergeant ordered a man to get his rifle and shoot near to the culprit. The soldier fired three shots as he was told, but with a fourth shot he hit the man. In the meantime, one of our Dayak collectors, Jagat, fetched a 12-bore gun, dashed into the jungle along the river (neither Kloss nor the sergeant seems to have made any effort to stop him) and killed the Papuan with a shot which must have been within a yard or two of the poor wretch. Nothing excuses the brutality of shooting a man for a small theft, nor the savagery of the man who gave this poor wretch the *coup de grâce*, and it is altogether a most miserable and deplorable business as it destroys our chance of making friends with the few Papuans we meet. I have no doubt that the mountain people, if there are any, will hear of this shooting (for news spreads even in a jungle), and our friendliness will be suspected. The Dutch themselves will be inclined to talk of the brutality of these Englishmen who come over here and butcher the natives; indeed it is a horrible business, for I am responsible for the acts of all the people in this expedition—barring the soldiers, etc. When I get back I shall probably have to attend some inquiry, and if the Dayak is sent to prison he will well deserve it.

OCTOBER 31. Left camp with Van de Water, Kloss and Dayaks, and went along Van der Bie's old track, which the Dayaks cleverly found yesterday. Fairly good going through open jungle, and then into foothills—direction north-south. Very straight; a bend to the south-west and west would save a lot of trouble, but to have a path at all saves us an infinity of work, for it would have taken us days to have cut the ground we walked over to-day. Camped on a steep hillside which V. de W. considers the worst camping ground he has ever used. He will have many worse before we go away.

NOVEMBER 1. Lost our path to-day at a river, and wandered about along ridges without finding it again. Camped on a high ridge. Lots of pitcher plants; very welcome, as these ridges are dry. Dayaks go well and cheerfully.

NOVEMBER 2. Struggled on through very bad ground and, reaching another ridge to the west, we found V. der B.'s track coming up from the south; thence along a high and steep ridge east to west where we found two old camping places and a house made by Meek, roofed with fan palms and still quite watertight after 2 years. We were all pretty well tired out.

The crest of the ridge is made of hard limestone—waterworn in a most curious way; much fossil coral.

NOVEMBER 3. Still along the same ridge, then down to a stream—the first we have touched for two days—where the Dayaks killed a large poisonous snake; they skinned it and kept the meat to eat. Uphill westwards, still following V. der B.'s path. Signs of pigs and also of natives. At last came to V. der B.'s old camp at the foot of Observation Hill; good camping ground about 2000 feet. Kloss and I went up the hill to Observation Point (about 600 feet higher up) and got a glorious view of the mountains; nearer ranges and east and west very fine. Lights and shadows not to be described. On the crest of the hill was another Meek house, roofed with leaves of fan palm and in a remarkably good state of repair. In it we found very clear evidence of recent visits of natives: bits of chewed sugar-cane, cut leaves not yet brown, remains of bones, and most interesting of all, two gourd penis cases exactly like those used by the Tápîro pygmies of the Mimika district. I believe the same type is used by the hill people of the Noord (Lorentz) river and the Island river. Round the house there was a clearing, and growing on it many sticks of sugar-cane, tobacco, ginger and bananas, all of which showed signs of having recently been tended. We sheltered from a heavy storm in the old house, and the Dayaks amused themselves by making fire with wood. I was much astonished to find that by using the 'string method', Kloss and I could make fire more quickly than they could. Found a native bow lying in the bushes, very much like the Tápîro bow. We decided to leave some stores behind us, so making the beginning of a depôt of food here. Hope it will not be taken away by Papuans. On our way down we saw far below us what appeared to be native houses. We went along the right track this time and camped at the spot where we had missed the path before. I believe another time the Dayaks could do this journey right through in one day, but I do not complain, as they have worked and walked well all day and seem to be happy. They are certainly the jolliest people to travel with I have ever met. Apart from what they carry for us, they carry a huge pile of luggage of their own, food for three days, cooking pots, clothes and all sorts of odds and ends, and *yet* the rattan basket—'ladong'—in which it is all held seems to be unfillable and never full! A great feature of these excursions has been V. de W.'s camp-stool which is carried by one of his convict coolies. Whenever we stop—as we frequently do—it is brought out for him to sit upon, and if he should want a bath in the stream, a man must carry it down for him to sit on while he washes—instead of sitting on the stones like an ordinary mortal.

NOVEMBER 7. Some Dayaks have just come up from base camp bringing food for themselves but none for the Dutch party. They say they will not bring any for the soldiers, neither will they carry any of the soldiers in their canoes, as the latter are so lazy and do not help in the work. I understand that a great many disagreements have occurred down at base camp, and Siddons, of course, has not enough grit to make the Dayaks obey him. It is a most confounded nuisance, and it is always the Dayaks from the rubber plantation who are at the bottom of any trouble. I wish I had not been compelled to take them, for they may infect my other Dayaks who so far have behaved very well.

NOVEMBER 8. Another worry to-day. One of our collectors and two other men went out yesterday to shoot, and they did not return in the evening. We fired shots after dark, but they didn't come, and I, for one, passed a rotten night remembering Stalker. However, this morning they have turned up, so all is well.

NOVEMBER 10. A heavy thunderstorm, ending in a furious hurricane, has swept a path a few yards wide right through our part of the camp. It has torn up small trees and flung branches about everywhere, spreading much havoc. Our kitchen has been absolutely demolished, and a falling tree has completely smashed part of the Dayaks' house; all their clothes and belongings are swamped, so naturally they don't want to start uphill to-day as I had arranged. Great nuisance.

I have been certain for a long time that we have made a mistake in coming to this Setekwa branch of the Utakwa instead of following the Utakwa proper. My original plan was to spend two or three days exploring in the motor-boat before fixing on a place for our base camp, but the Dutch people made such a point of the necessity of the *Valk* returning to Ambon as quickly as possible that I had to drop the idea, make use of Van der Bie's old base camp, and consequently to follow on his route towards the mountains. I cannot but believe that if we had chosen a place on the Utakwa river for our base camp instead of on this branch of it, we would have taken less than six days to reach that spot where we hope to get to the Utakwa valley. But there is one advantage, and that is that V. der Bie's cut track has saved us an immensity of labour.

NOVEMBER 13. A very sudden and furious flood last night, and at daylight this morning we found all our canoes had been carried away. Managed to get back four, and the Dutch got back three of theirs, but only two or three are fit for anything. We are of course now very badly off for means of transport, and it is most unlucky that this big flood should have come upon us on the one day when all our boats were up here.

NOVEMBER 16. I have sent fifty-three men carrying loads up to the First Depôt Camp, and I am now down at Base Camp. Have seen fragments of our broken canoes lying about in the water. It is really confounded bad luck.

NOVEMBER 19. (*Base Camp. To his Mother.*) All goes well, and the health of the people so far is excellent; cut feet, stomach-aches and mild attacks of fever are the worst. We have made a large camp at a spot two days by canoe from here, and we are making another store camp at a place some 2000 feet up (a three days' march from the river camp). As I told you before, my plan is to go on making these store camps further and further up, and from the last camp I hope to attempt to get to the top of the mountains, or as near to the top of them as I can. We have had a good view for several miles up the valley that we propose to follow. It appears to lead directly to the Snow Mountains, but it is desperately narrow, and it may turn out to be so steep that we shall be unable to follow it for any great distance. In the early morning we can sometimes get a magnificent view of the snows, only about 30 miles away, and I have seen them well enough to be sure that if we can reach the foot of the snow we can also reach the top. If we can look over on to the other side, and see further and perhaps higher ranges beyond—no man knows—it will be worth all the grief and pain that lies on this side, for it is going to be the devil of a job at the best. Everything depends on the Dayaks. If they stand the cool climate and the unaccustomed rocks I believe we shall have a fair chance of success. They are splendid fellows, especially the forty-two men I brought myself from their homes. I begin to have quite an affection for them. There is such a lot to do, and these next four months are all too short for the work we want to get through.

NOVEMBER 22. (*Base Camp.*) Have begun to make a map of the mountains and have taken some telephotographs; not absolute failures. Shall learn how to do this in time.

Some Dayaks have just brought me a string bag got from some Neweripa river natives. So far as I can see it is exactly of the same pattern as that used by the Tápîro pygmies.

NOVEMBER 26. (*Base Camp.*) Very hot day; 89° F. in shade (i.e. inside our big storehouse—the coolest place in the camp). Another tremendous thunderstorm and a furious wind which carried off many roofs in camp.

DECEMBER 1. (*Base Camp.*) Tedious business waiting for the boat with stores, due more than a week ago. At midday a canoe was seen coming up the river with the Dutch flag in the stern and a man sitting under an umbrella in the middle. We thought it must mean that the launch of the Dutch boat had broken down, and that the natives had given them a lift up river, but found to our great disappointment that it was only our old friend the 'Maioor' of Nimé, who had come to pay us a visit. Very glad to see each other but he was a poor substitute for the long-expected boat. Soon afterwards some other canoes full of Nimé people came along and we did a great trade in bananas and coconuts and ethnographical odds and ends. Many of the men had familiar faces and they recognized me. They had come all the way by sea from Nimé, slept at Poeriri, and so did 40 miles by water to-day. They did not stop at the Neweripa village though they said they saw it, so I take it they are not on visiting terms with the people there. The Maioor promised me a great feast of pigs and bananas should I go to visit him at Nimé. Hope to be able to manage it in March.

DECEMBER 7. Still waiting for the wretched steamer. Have now lost fifteen days uselessly....Much trouble has been caused by the confounded people at Batavia who did not pack my rice in the way they were told to, so that a great deal was lost and much spoilt. Another thing is that the Dayaks, instead of eating eight kilos of rice a day as I was assured that they did, actually eat one kilo—a very considerable difference. They also eat twice as much meat as I was told they would, and other things in proportion, so I have had to order a quantity of new stores which is just what I had hoped to avoid. Nobody knows anything about the feeding of these Dayaks, even those who said they must have only pig-meat were wrong. I have been feeding them on dried pig-meat at 4.46 guilders a kilo when I might have given them dried cow-meat at half the price—an experience very dearly bought. Otherwise my stores have worked out very well and they are all good. My forty-two men from Bakoel and Landjak are excellent in every way, but those from the rubber estate and the eight Sarawak men are more trouble than they are worth; the best thing they have done has been to make a canoe which stands knocking about the river better than any of the others.

DECEMBER 8. Up river to 'Canoe Camp'.

At this point in the diary A.F.R. makes the following note: 'On March 9, 1913, coming down the river from Canoe Camp to Base Camp, my canoe upset in a dangerous rapid, so that I was nearly drowned, and I lost the greater part of my baggage, cameras, medicine chest, maps and diaries. Now I have to begin to write over again my diary from December 8, to March 9—so far as I can remember it. A.F.R.W. March 19, 1913'.

DECEMBER 14–18. Up the river again and spent three days at No. 3 Camp, building ourselves a house of palm leaves and generally making camp good. Afterwards left this camp and made our way towards the mountains. We took with us two or three hill Papuans and they were very useful in showing us a track. Made good progress. Came to a fair-sized river in a pretty gorge—bed of boulders—then over low spurs passed a curious native shelter under a huge boulder; down to another fair-sized tributary of the Utakwa—a milky river about 30 yards across—then up and along a very steep and almost knife-edged ridge from which we heard the roar of the Utakwa on our left and of the milky river on our right. Followed the ridge, then down, and camped about 200 yards from the Utakwa; many fan palms, so it was easy to make a good camp.

Next day, led by Papuans (general direction nearly due north), we walked until we got to a smallish river coming from the north-east. Met some natives coming from the mountains with women and children and several pigs. They seemed very anxious to prevent us from going further, and after a great deal of talking an active old man—whom we afterwards found to be a sort of chief among them—took a gourd of water and waved it about, scattering the water into the stream. He ended by throwing a stone into the middle of the stream and shouting something at the same time. After informing us that we might go on, he set off himself in front to show us the way. We followed the stream for about a mile—in two places wading up it through a cañon with steep walls of rock, and in other places running over stones and boulders—until suddenly the old man left the river in a most unexpected way by climbing up a steep waterfall that came down from our left. We struggled up a hill to about 4000 feet—on through broken ground to the top—and then steeply down again to the Utakwa. Passed another rock shelter, with hill natives, pigs and children, etc., camped here.

DECEMBER 19–20. To-day we got to a place where a large landslip had carried away all the trees and vegetation, leaving a steep slope of screes, about 250 feet or more in height. We climbed slowly and diagonally up and across this to its top, and scrambled out on to firmer ground. Then up and down very steep spurs close to the river. Now and again we could follow the river itself, but generally the banks were too steep—hence this up-and-down business. After a few hours we came to the junction of the Kemarong river from the west with the Utakwa. Here we found that the natives had built a bridge, a very rough affair of logs thrown from a boulder about a third across to the other side. This bridge was quite new and evidently made by the large crowd of natives who had all been coming down these last days to visit us at camp. A few yards from it was the usual native rattan rope bridge, stretched across the river from posts wedged into rocks on either side—about 40 yards long and about 20 feet above the water. Hanging on the rope were rattan rings about 2 feet in diameter. The native's method is to hold the rope with both hands, thrust his legs to the knees through one of the rings, and so—hanging parallel to the rope—slide down to the middle, and with hard pulling 'uphill' (i.e. of the slope of the rattan) work up to the other side.

Made our Second Depôt Camp on the ground between the two rivers. We had intended to go back to our First Depôt, but the natives with us—now increased to thirty or forty—were all so anxious to go on up to the 'Ingki-pulu'—as they call the Snow Mountains—that we decided to go on, taking five Dayaks and food for six days (which was all we had).

DECEMBER 21. Guided by natives we went on north, up and down many steep spurs running down to the Utakwa, and in about four hours we came to a newly cleared piece of ground in the middle of which was a house—apparently quite new. It was the first native house we had seen, for all those we had hitherto seen had been merely wayside shelters —not permanent houses. This one, typical of those seen afterwards, was built on piles about 6 feet from the ground and about 10 feet square; entrance by sloping planks or logs to an outside platform leading to a square room, in the middle of which was a fireplace. Outside the house were a number of men, women and children, mostly occupied in preparing a very savoury-smelling feast of sweet potatoes, yams and pig flesh. The cooking was done in round pits, about 2 feet deep and 3 to 4 feet across, into which they put hot stones wrapped in leaves, then potatoes wrapped in leaves, and more hot stones on top. The people cooking the feast made mysterious signs and pointed towards the house. In particular they made the curious sign of lifting up the right nostril with the right thumb, which we afterwards came to associate with death. When we went into the house we found there a miserable old man, very sick and covered with horrible sores, with four or five other people about him. He seemed to be glad to see us and the people appeared to expect us to do something, but I could not quite make out what. When we returned to this place four days later we found the house burnt to ashes, and lying on the top of the ashes were the remains of human bones—doubtless those of the sick old man. The puzzle was, whether he had suddenly and unexpectedly died and they had cremated his body, or whether they had known he was going to die. This can hardly have been the case, for he was by no means dying when I saw him. Perhaps the feast that was being prepared was a funeral feast, and they set fire to the house at the critical moment; or perhaps they knocked him on the head first; anyhow they utterly destroyed a fine new house and then deserted the place altogether.

We camped in a very beautiful spot looking north, and in the evening we had a fine view of the snows of Carstensz by moonlight.

DECEMBER 22. Long dispute among the natives as to which way we shall go. There seemed to be two possible routes from here, an east and a west; final victory of east. We went steeply downhill to a big branch of the Utakwa (Nusula Narong), which we crossed by a very insecure bridge of three saplings between two immense boulders; then onto a knife-edged ridge between the Utakwa (here called Singgarong) on our right, and the Nusula Narong on our left. Following the ridge and passing a sort of rude fence across the path, we found ourselves on a levelled platform about the size of a lawn tennis court; ground quite hard and dry. Here were about sixty or seventy people of all ages and sizes, who set up the most extraordinary barking when we appeared, dancing and prancing and waving their bows and arrows.

Some came and shook hands, or rather pulled knuckles with us. You hold out the bent knuckle of your middle right finger, while the other person grips it between the bent first and second fingers of his right hand; then you both pull until your hand comes away and the other one's two knuckles come together with a click. This was repeated three or four times, accompanied by a 'Wah' or other ejaculation. We were then told to stand with our party at one end of the platform while all the natives belonging to the place stood at the other. A man, a sort of boss among them, ordered silence, and then began a long harangue. In one hand he held a rough iron axe, in the other two white leaves, and towards the end of his speech a lean white pig was brought in from the back of the crowd and I was instructed to go forward and receive it, which I did. Rather an embarrassing gift, but happily I was then presented with a small boy and girl as guardians of the pig. We gave them a small present, and then were told that we might proceed through the country of these mountain natives.

We followed along the ridge, climbing to about 5000 feet, and on our way saw many extensive clearings on the slopes of Mount Venus, up to 6000 feet or more. This ridge was very steep, and although covered with dense jungle, quite waterless. We camped on the crest of the ridge—the only possible place—and all that afternoon we were visited by natives crowding about us to suffocation. They pull you about, poke you, stroke you, pull your hair and beard, and generally fondle you in a way to which I am quite unaccustomed.

Very cold at night (about 60° F.).

DECEMBER 23. On again up and along the ridge, striking off to the right—a little north-east—till we came to a collection of four huts where people were standing waiting for us. Over another spur, down a steep slope to within a short distance of the Utakwa, and then onto a level spot where we camped beside a bubbling spring of water; very good. Here again there was a native clearing with sweet potatoes and sugar cultivation. When the slopes are steep, as these were, the earth is terraced and kept in position by logs. Again we had much knuckle pulling to do, and we were stroked and cuddled more than ever before. All these last days we have been accompanied by natives, who when they come to within shouting distance of clearings, all start shouting in various keys; some a deep 'Wah', some a high sort of yodelling note, and others according to the compass of their voices—but each one a single note only. The effect is very pleasing, and when you hear it at a distance—as from the top of a ridge or the bottom of a valley—it reminds you uncommonly of the cry of a pack of hounds.

Late in the afternoon we heard loud shouting from a hill in the direction of the higher mountains. Soon the shouting turned to the more regular barking, and in a few minutes about forty men and boys, wearing their cassowary busbies and waving bows and arrows, came running in single file towards us. When they got near they formed themselves into a sort of wheel-shaped figure running round and round. Afterwards they mixed with the others and began to visit us. We were quite exhausted by six o'clock, when the last of our visitors left, and I realized something of the social hardships of monarchs.

We decided not to continue further owing to lack of food, so we returned to the camp of the 21st. It was a very long way, and for me a very painful way, as I wrenched my foot badly soon after starting.

DECEMBER 25. To-day we ate our Christmas dinner under a single very leaky tent fly. We had a tin of beef and rice and a Crosse and Blackwell plum pudding; two guttering candles, and streams of water underfoot. Pretty dismal.

DECEMBER 28. Two of our Dayak canoes had trouble with some coast natives, ending in the Dayaks offering no resistance and running away. The canoes were utterly looted, and we lost all our chocolate etc., woollen caps, shirts, sweaters, blankets, and to my great annoyance half a pair of new boots of mine—specially kept for the mountains. Altogether a desperate loss, and I had now to go down to Base Camp to see how to make up for some of these things.

DECEMBER 29. Walked down to No. 3 Camp and did it in 9½ hours. Dog tired, but drank a bottle of Lord Rothschild's champagne and blessed him.

JANUARY 2–23, 1913. Went up river for the fourth and I hope last time. Started for the Snow Mountains again with Kloss and Van de W. and fifty-six Dayaks. Camped at the same places as before, and as usual we were accompanied by natives who seemed to enjoy carrying the men's loads for them. Left two collectors on the ridge near the burnt house (about 4200 feet), and camped at the place of December 31—which was a pleasant camp giving a fine view of mountains. Next day had a long march to the place we reached before in two days, and this time found the natives not a bit excited about us; they hardly troubled to come and look at us and we got no vociferous welcome as on the first journey.

JANUARY 24. Started through new country. Crossed a ridge, and came suddenly upon a glorious view of Carstensz—now less than 10 miles distant. Luckily the natives appear to have a track up to the very mountain or we might waste weeks in looking for a route. Signs of very extensive cultivation in the first valley and also on the ridge beyond, but beyond that we saw no more cultivation, and the natives tell us that there are no people living further up. These natives are very greedy, and when we offer them axes for their pigs they say they must have two perangs for each pig, so we do not trade much with them. They talked a great deal about 'piu', i.e. salt, and asked us to go with them to see the 'salt place'. We went, and after going through a great deal of bush came, not to a deposit of salt as we had expected, but to a warm sulphur spring; milky white water, gently steaming, 94° F.; with a very strong smell of sulphur and a nitrous smell of some sort. The water oozes out of the ground in several places, and in one they have dammed it up to make a round pool about 15 feet across and 2 to 3 feet deep. The man and woman with us took stems of the long grass that grows near the pool and proceeded to suck up the water as through straws. They drank a great deal and appeared to like it, but we tasted the stuff and found it very nasty. When they had drunk enough, the woman waded into the pool—stirring up many bubbles and an increased stink—and groping along the bottom brought up armfuls of stems and leaves of ferns and other shrubby plants which had evidently been left there to soak; these she put in her bag and carried away. Evidently the people chew these stems that have been soaked and find much good from them.

At this camp there were violets and many different kinds of begonias.

JANUARY 26. Up the valley again with half a dozen natives; down to the river and over very steep spurs—so steep that in places we had to tie ropes of rattan to let ourselves down by—so the going was slow with a long line of laden men. The limestone rocks of which the ridges are made slope at a high angle down to the river hundreds of feet below. We had to cut our way through a scrubby kind of bush 10 to 20 feet high, held together by a sort of blanket of moss and orchids. Sometimes the weight of men made the vegetation begin to slip downwards over the smooth rock, and not very much more weight would have started the whole thing avalanching down into the torrent. In one place the bank was too precipitous for us to go up. There were no footholes at all, so we had to make a kind of bridge along the edge with a long sapling thrown from boulder to boulder. As we walked along this narrow sapling, we could touch the rocky wall with our fingers, though there was no crack or projection to give any handhold in case of a slip—which would have meant being swept away into the torrent like a piece of grass. I was very glad when we were past. Camped in a mossy jungle and slept snug enough in the Mummery tent.

JANUARY 27. To-day we started by wading through very cold streams, then went steeply uphill where the character of the country changed rapidly. River widened, trees became smaller and less dense. Casuarinas and coniferous-looking trees. Flowers much more numerous: orchids, a sort of meadowsweet, a rose-like bush and many others. No track, but natives have evidently been here before and we saw signs of former fires. Camped beside a boulder in the middle of the river bed; not a safe place if it rained heavily but fortunately it did not. (This is Camp No. 11, 8000 feet.)

JANUARY 28. On up the valley, and the river now easily fordable—just a mountain stream tumbling over great boulders. Here I picked a small blue gentian and saw a pair of very pipit-like birds. At about 8500 feet are undoubted glaciated rocks. At 9000 feet a thin band of a sort of brown coal. The stream now came to an end, or, I should say, it began. A little above 9000 feet it trickled out of limestone on the top of clay. We got on to a sharp-edged mossy ridge (a spur of the main mountain mass), covered with a very weird vegetation: 'pine-like' trees, bushy heaths and rhododendrons (?), with many pretty flowers. Along the edge of this ridge was what appeared to be a well-worn track like a much used rabbit run, and the natives said that it was made by an animal called 'op'. They showed me a place where the animal had pecked curious holes in the ground and I feel sure the animal is the Proechidna. It must be extremely common here, but we cannot persuade the natives to catch one. At about 10,500 feet, I saw by the side of a little stream the droppings of a 'game' bird, probably the so-called 'partridge' that Lorentz found; bird not seen. On along the ridge—mostly in clouds—and then we camped. The natives did not quite seem to know where we were. No water, so we had to make our tea with moss water until it rained in the afternoon. When the clouds lifted, we found that we were right up against East Carstensz, which towered over us though the top was not to be seen. The edge of the ice-cap of the mountain seemed to hang almost over our heads, 4000 feet above us.

JANUARY 29–30. Camped in a wide hollow with plenty of bushes to put down for bedding.

V. de W. and I, with three natives and a few Dayaks, started off to find a way of approaching the snow. Very bad going through dense moss forest. One thing is certain, and that is that without the help of natives we should never have found our way through this very difficult country. Got on to a little col between the main mountain and an outlying spur. Magnificent view back to all the mountains we had passed, the whole of the Utakwa valley, the gap in the trees at Observation Point, and the flat plain of jungle with rivers winding away to the sea. Three rivers visible to the east of the Utakwa, and to the west of it—all the big rivers as far as the Wania. A most notable view. We climbed on over nasty loose screes, and then on to more open rocks with a sharp needly ridge on our left and a big precipice on our right. A sudden wall of rock appeared to block our way, but one of the natives managed to get up it followed by a Dayak carrying the Alpine rope (the same piece I took to N.G. in 1910 and never used). The rope was fastened to a rock about 100 feet up, and without it this place would have been really difficult and dangerous. We scrambled along some limestone rock until we found ourselves on a small patch of snow—a sheltered hollow between steep walls of rock. Above this was a large sheet of snow or rather ice—a kind of dribble from the ice-cap proper. Looking straight up we saw the seracs of the glacier, and scattered about us were pieces of the ice that had broken off. Time midday, and it was out of the question to go up, for it was much too steep for two people, one of whom had never been on ice in his life. We decided to skirt along the edge of the ice westwards and look for a possible way of ascent. V. de W. was in tremendous glee at being the first white man to reach the snow of Carstensz, but the thing is to reach the ridge. Went back to camp well pleased with ourselves.

JANUARY 31. Kloss and I with two tents, one for ourselves and one for Dayaks, went off to make a camp as high up as possible so as to get higher on the morrow. Horrible morning, raining and blowing so that the Dayaks were very wretched with cold. Many of them threw down their loads and turned tail in spite of the most horrible threats of wage cutting, etc. At 12,000 feet we could get them no farther, so stopped and camped; dense mist and everything soaking wet. The rain stopped at midday, and at once there was no water at all as this porous limestone absorbs every drop at once. All of us too tired to go down 800 feet to the nearest water, so we had only half a water-bottle full to last until next day; I don't think the Dayaks had any at all. We lay on a steep slope of very lumpy ground with holes filled as much as possible with cut bushes; underneath us was the Mummery tent and over us two blankets; extraordinarily uncomfortable, but we had an exceedingly fine dinner off a tin of mutton which down below had always been sloppy tasteless stuff—now it was hard and firm and excellently tasty. Also half a tin of plum pudding with a dash of brandy from my flask, and then all our water in cocoa, which gave us a small cupful each. It was all very good had we not been so tremendously hungry and capable of eating ten times the quantity. The Dayaks kept a fire going most of the time in their tent so they kept a little warm, but nobody got much sleep.

FEBRUARY 1. This morning there were clouds and fog and occasional rain, but we got to the foot of the snow and worked along one of the terraces of rock to the west for half a mile or so, and then upwards over rocks to another ice overflow. We climbed rocks on the side of this for 300 or 400 feet and then found ourselves faced by the choice of a steep rock climb or an even steeper ice climb (i.e. up the bergschrund of the ice-cap). There is apparently no easy snow way up to the ridge from here. If we had been three men accustomed to ice we could have gone up, but it was not to be thought of for two, of whom one (Kloss) had never been on snow in his life and knows nothing of climbing. Nothing for it but to turn back. Boiled hypsometer—14,866 feet. Horribly disappointing, but the weather was very thick, and if we had got to the top of the ridge we should have seen nothing. The lowest point in the ridge between East and West Carstensz could not have been more than 500 feet above the point where we stood. We left our Alpine rope to assist the next party that comes here, and arrived back at camp wet through.

So far as I can make out, this mass of Carstensz is a limestone ridge, running west by north and east by south, tilted at angles from 20° to 60°. The south face is so steep that the snow does not lie on it except in the hollower places. Snow lies on the top, and forms an ice-cap which flows over to the south for a few hundred feet only and then breaks off in the sort of bergschrund mentioned before. Where the snow is less steep—in troughs as it were—the ice flows further down, making miniature glaciers crevassed in the usual way. It is fairly safe to assume that the slope is less on the north side, and that the extent of snow and ice is greater. On the south we found snow at 14,200 feet, but the lower edge of the ice-cap is about 200 feet higher and ends abruptly.

Nobody knows how terribly disappointing it was to have to turn our backs on all these questions, which were what originally brought me to this country three years ago and which still remain to be solved by the happy person who first reaches the watershed and looks over to the other side. By getting to the snow we have accomplished nothing of any value whatever, whereas if we had gone a few hundred feet higher and had seen beyond, what might we not have seen? Further and further ridges? This prize was withheld—and nobody knows.

We were compelled to go back, our food was at an end, and men all wretched with the cold. Fires won't burn properly on account of the damp wood, and rice won't cook properly because of the low boiling-point of water. Stomachs are out of order as well as spirits.

FEBRUARY 8–10. Travelled down to Depôt Camp, and half-way down a very steep ridge we came upon the body of a man who had died apparently four or five days earlier. Half a mile further was the body of an oldish woman lying in a small stream, her bag and belongings lying on the track a few yards away. The natives who were walking with me made me understand that they had died of hunger, and when we got to Depôt Camp we heard that a number more of these poor natives had died in the same way further down.

From No. 6 Camp down to No. 3 Camp was a walk that I shall remember as long as I live. Soon after starting we came on the body of a man not long dead, then the bodies of two women, one child, and another man. On further, many more bodies lying in ones and twos—some dead in the track—some in rock shelters, and some in roughly made huts of leaves; one or two had been buried, but most were left just where they lay. All this had happened in the last three weeks, for these bodies were along the same ground we had passed three weeks earlier. In a leaf hut Kloss found a little girl of about two years, alive; a man, woman and child lay dead beside her. Perhaps the poor little creature did not belong to them, but had been left there by people who had not wanted or had not had the strength to carry her further. Kloss was alone at the time and he carried the child for six hours in one of the net bags he found lying in the hut. We fed her on milk and cornflour and though she was very feeble she was talkative, and I did not think that she was in a desperate state; however she died in the night. The whole business is very distressing and very hard to explain. These people come down from the hills with their women and children and pigs to see us in our camps; they bring bags full of sweet potatoes, and all goes well until the food gives out. They then beg food from us and we give them rice, but not a great deal, as our amounts are calculated for our own men and we have no large surplus. They cut down palms and eat the growing heart without cooking it—and doubtless many other unaccustomed roots and fruits—and then they begin to think of returning to their homes in the mountains. Already weakened by underfeeding, hampered by women and children and often by small pigs, they have a long journey of many days up and down hill through an absolutely foodless country before they reach their potatoes in the mountains. Altogether we reckon that not less than forty people have died of those who have come down from the mountains to see us, and the only cause that I can see for their deaths is starvation. It is very puzzling, for they are not greatly emaciated. I believe they just give up and die as natives can do. Those that were buried were tied with pieces of pandanus leaf into a sitting position and lowered into a round hole, 3 to 4 feet deep. The hole is filled up, a few tall sticks stuck into the ground about it, and one or two of the dead person's possessions hang from the sticks.

FEBRUARY 11–MARCH 2. We are resting our weary bodies, filling our empty bellies, sleeping, reading newspapers, taking photographs. Am also improving my map; altogether quite a pleasant time. Started sending things down the river and decided to follow in a canoe laden only with my personal belongings.

MARCH 9. River rather full and canoe bad and small. Four miles down, in the worst and deepest of the rapids, we touched a sunken tree, upset, and the canoe went careering bottom up down the stream. A Dayak and I clambered on to it, went along for 100 yards or so until we approached one of those big fallen trees lying over from the bank into the water. The Dayak sprang away from the canoe telling me to do the same, but I was hampered with clothes and not a great swimmer at the best; so I got caught in some of the underwater branches of the tree, was dragged deep down to where it was horrible and quite dark, and after a desperate struggle I freed myself and came up, half dead. I was then carried along by the current at a hideous pace, now and again touching the ground with my feet, but utterly helpless. After nearly half a mile I found myself in a strong rapid of about 2 feet deep, running over large stones. I was too much exhausted to stand up, and my boots (moccasins) were too slippery to grip to the stones. I half lay down, but the water in my clothes dragged me along, so I kept in a sort of half-kneeling crouching position. In this manner I was slowly dragged along the stream towards another swift and deep rapid, which would certainly have been the end of me, had not the Dayak caught sight of me, clutched me out of the water and dragged me to the bank where I felt pretty bad for some time. The canoe was recovered, but the sack containing my bedding and everything else was lost.

Plane table and map.
All the instruments of the R.G.S.
Telescopic alidade.
Max. and Min. thermometers. } Subsequently recovered.
Prismatic compass.
Aneroid and hypsometer with three thermometers (R.G.S.).
My own aneroid.
Sanderson camera with Zeiss lens, Zeiss telephoto lens and six dark slides.

My new compass—a very good one.

My Congo medicine-chest, used on Shackleton's Antarctic expedition, and by me on the Mimika expedition.

Diary from December 8 onwards.

Folding chair and table.

Silver flask.

Old silver pencase and pencilcase.

Many books and many other things I can't remember. Worst of all of course is the loss of the map, of which I can only make a very sketchy business now. Then my diary—It contained all sorts of observations of the natives that I cannot possibly remember. I am left utterly without medicines, except quinine and castor oil. Desperate and hopeless condition. Within an ace of being drowned into the bargain, so I ought not to complain too bitterly, but it is very hard to be philosophic in the circumstances.

At Base Camp A.F.R. spent a few days exploring the mouth of the Imabuka River. There was some difficulty in getting over the bar, but they managed it and anchored about a mile inside.

Here we were visited by natives in canoes who declared that there were no people at all living up this river. Along the coast to the Koopera Pookwa River (the natives called it Kooperoópoowa), where we found a big native encampment and were greeted with dancing, but nothing so striking as the Mimika welcome. About six hours up this river we came to a long village, much the same type as Wakatimi but for the time being quite deserted. I don't think anybody has ever been into this river before us.

APRIL 3. Sailed from New Guinea. I shall be very sorry to say goodbye to many of my Dayaks. I have brought all away except one. They are nearly all fit but pretty well fagged out. We are very thin and I should say that between us all we must have lost nearly a ton of flesh. I hope to make up some of mine on the voyage home.

A.F.R.'s Notes on the Mountain Natives

It is hard to say how many people we saw, but I should say that at one time and another we saw altogether about 400 individuals, perhaps not so many. From the beginning they were always very friendly, coming of their own accord down to our camps and bringing their wives and children. They were quite fearless and would march into our houses or tents as if they had been accustomed to do so all their lives. They are small*ish* people, but noticeably taller than the Tápîro, and we saw a number of men who were certainly 5 feet 8 inches and upwards. They are strong and well built, buttocks and calves finely developed, and they run and walk up and down hill very actively. At first they appear to be almost black, but most of this is due to the filthy condition of their skins, caked with dirt and sweat. Some of those who came down to Canoe Camp shaved their heads and soaped themselves many times a day, with the result that their skins showed almost yellow. The men like to paint the upper half of their face red with clay, or black with fat and charcoal. Many of them have a very uniform growth of hair all over the body and limbs; hair on the head short and woolly, not dressed at all; whiskers and short beard frequent. The septum of the nose is pierced and sometimes a pig tusk is worn in it; alae nasi not pierced; lobe of ear pierced for ornaments frequently worn; no tattooing nor scarification. The dress of the men is the long gourd penis case, almost exactly similar to that of the Tápîro, kept in position by a string round the waist. Hanging over it is a double tassel of knotted string, something like a sporran. They have a curious habit of clicking the penis case to show astonishment. The women wear a short petticoat of bark cloth which always seems to be just slipping off behind. All—men, women and children—carry haversacks of netted string—purse stitch. They sometimes have a big one slung across the shoulders, and a small one—a sort of treasure bag—hanging round the neck; in the big bag nearly all their worldly goods are carried: their pandanus mat, tobacco, fire stick, food, and all sorts of odds and ends. I have seen a woman carry in her bag—in addition to the usual stuff—a baby, a young pig and at least 100 sweet potatoes. Both men and women wear necklaces, made of either seeds, beads, teeth of pigs, bones of rats or bits of broken shell. Another common ornament is a sort of fillet of the upper arm bones of a small mammal, worn in the hair. The most imposing of all decorations is undoubtedly the 'busby' headdress of cassowary feathers fastened on to a hoop of wood; it gives the wearer a very fierce and warlike appearance.

We saw altogether four or five metal axes; origin unknown. The usual weapons are bows and arrows—the latter of several kinds—each with a distinguishing name, and much better made and finished than those of the coast Papuans; so are their stone axes sharper and harder than those of the coast people; they are set in a wooden and not bamboo shaft. Nearly all carry a stone knife, a straight or slightly sickle-shaped flake of a slatey stone about 1 inch wide and 4 to 8 inches long—capable of taking a very sharp edge. Some carry a spoon made from the shoulder blade of a pig or other animal, and used for scooping out the inside of cooked potatoes.

There is an animal that lives in the mountains which they call 'op', but in spite of the offer of a large reward they never caught one for us.

They call the Snow Mountains 'Ingki-pulu', but we cannot discover what they call themselves. They seem to live between 4000 and 6000 feet high, but their hunting tracks go much higher, occasionally up to the snow. Besides potatoes they grow yams, sugar-cane and a few bananas, but potatoes are their chief food. No intoxicating drink as far as I know. Tobacco in some quantities (arenyum); the leaves rolled into a long rope then coiled into a neat twist; flavour not at all bad. They made cigarettes, using pandanus as wrapper, and sometimes they smoke from the side—if the wrapper is not wide enough to go all the way round. They also smoke pipes made from plain straight bamboo with a hole at the bottom. Women and children all smoke.

No signs whatever of cannibalism. When anyone is very sick or dead, or if they are talking about something unpleasant—as for instance about the snow at the top of the Ingki-pulu—they have a curious habit of lifting up the right nostril with the right thumb and making an expression of disgust. Beside the finger-pulling described before, they greet you by raising the brow two or three times and nodding the head to the side once—smiling at the same time; many have very agreeable smiles. We used to try and do likewise with our brows, but soon found our foreheads becoming quite tired with the exercise when a large crowd of people were visiting us. Their voices are loud and harsh and they talk incessantly. We have made a considerable vocabulary, but of course only of the names of things. They have words for numbers up to ten, and by using fingers and toes they seem to be able to reckon fairly accurately up to twenty.

Men seem to be considerably in the majority and I do not think that one man has more than one wife. I noticed a very large number of men and boys who were blind in one eye. Whether it was the result of accident or intention, I do not know. On the whole they seem splendidly healthy, but I saw one or two people who had undoubtedly syphilis.

We saw no carved images, no fetish houses, and no signs of any 'religion' at all. Nor did we see them play any games, but they are familiar with cat's cradle. They have Jews' harps, but no drums or wind instruments.

For this expedition and for his part in former expeditions, A.F.R. was awarded the Gill Memorial of the Royal Geographical Society, and in 1925 he received the Patron's Gold Medal of the same Society.

In March 1920, a Paper was read before the Royal Geographical Society by Mr E. W. Pearson Chinnery, entitled 'The Opening of New Territories in Papua'. Following the lecture, A.F.R. made these remarks:

It is very nice to see the world opened up: we like to see new lines of railways on our maps and to see Africa dotted with aerodromes and such things, but I do not think we sufficiently consider the point of view of the people whose countries are opened up. Mr Chinnery objects to inter-tribal warfare. Well, we have spent years in killing each other, at great expense, to make the world free for democracy. Now and again the Papuans kill one or two people to celebrate a festival, or perhaps because the country does not produce beef and mutton. I do not think really it is quite fair of us to inflict what we are pleased to call our Western civilization on these people. You call them 'savages'. Many of these people—not those you have seen photographs of to-night, but their cousins who live a few hundred miles to the west—are personal friends of mine. I have always found them to be a happy and cheerful people, sufficiently fed and suitably clad. So far as I know they are as truthful as most of us, and in many months I have spent with them, though they have had endless opportunities and unspeakable temptation, I have rarely known one of them to steal. The lecturer says we must alter—modify—their traditions ('institutions', I think, is the word) so that they may 'fall into line with the needs of progress'. I hope they will go very slowly about this modifying of institutions. You have in New Guinea the last people, I believe, who have not yet been contaminated—if that is not an unkind word—by association with the white races. They have an extraordinarily interesting culture of which we know very little, and we have much to learn from them. I suppose it is too much to expect that the whole of the interior of New Guinea should be kept as a vast ethnological museum; but I should like to believe that the Australian Government will set apart a really large area—there is plenty of room—to be kept as a native reserve where these people can live their own life, and work out their own destiny, whatever it may be. Into that country no traders, no missionaries, no exploiters, not even Government police themselves should be allowed to go. There are, of course, difficulties, but they are not insuperable. The inland regions are invariably difficult of access, so that it should be easy to prevent an invasion of undesirable outsiders. The mountain tribes wander very little from their own valleys, and if approach from the coast were cut off they would live undisturbing and undisturbed by their more sophisticated neighbours. Perhaps it is an impossible dream, but I am looking ahead through two or three or more centuries, and the example of the fate of the Tasmanians and the present condition of the aboriginal Australian natives ought to be a sufficient warning.

VII

THE WAR

When in 1914 war broke out, A.F.R. was plann ng a third expedition to New Guinea. In his earlier war diaries it will be seen that he sometimes writes as though he were telling his experiences to a friend. The explanation lies in a letter to Sir Henry Newbolt, written in 1919, when he says: 'I am amusing myself at odd moments in writing letters of a naval surgeon '14–'19, but I doubt if they will ever get beyond the pencil copy'. I have only been able to find these 'unaddressed letters' up to December 1915, when he left the *Agincourt* for East Africa, and I have thought best to place them as though they were part of his diary.

On August 1 I went home with the intention of staying a few days, but on August 3 I heard that Great Britain certainly goes to war. I returned to London, dined at the Savile, and there found myself opposite to a man who told me that the War Office is already sending out notices asking for medicals. He held forth at great length about the needs of the Army in the matter of doctors. I suggested that the Navy was pretty short of them too, and that the Navy would probably have need of them before the Army began to get moving. Anyhow the man bored me so much with his harangue about the Army that I determined to throw in my lot with the Navy—if they would have me.

On the 4th I went down to the Admiralty and was interviewed by a courteous official who informed me that a few months later I should have been above the age limit. I don't mind betting that I am a good deal tougher than many of the lads straight from college or hospital. The medical examination was quite thorough, and though I felt a perfect fool hopping up and down a room stark naked, I think I surprised them by the way I could climb a rope. In

December '99, when I wanted to go to South Africa, I was rejected as being medically unfit for the Army. I think I have probably had far harder times since then than anyone had in the Boer War. However, the result of my examination is that I am now called a Temp. Surg. R.N., and barring incompetence or ill-health I am liable to serve for the duration of the war—or a period not exceeding five years. Great bore all this, for I was to go off to Scotland last night to spend a fortnight's fishing in Argyllshire, but the War Lord—or whoever is at the bottom of this business—has decided otherwise.

Feared that I might be sent to a shore hospital, so went to the Admiralty and got Herbert Richmond* to support my application to be sent afloat. While waiting at the Admiralty I watched great numbers of people going in and out on urgent business, all very much occupied but very orderly and no sort of fuss at all; even the chief hall porter found time to come and give me his views on the war and on the Germans.

AUGUST 7. Back to my rooms where I found a telegram from the Admiralty ordering me to join the S.S. *Mantua* at Tilbury Dock, 'forthwith'. That last word sounded rather urgent, so I packed up, wrote letters until 3 a.m., and at 5 a.m. caught a train to Tilbury where I arrived before breakfast. I had forgotten that the Navy is either shaven clean or bearded, so I clipped off my moustaches with a pair of scissors in the train from Fenchurch Street. The other occupant of the carriage probably thought I was a German trying to disguise myself. After all it appeared that 'forthwith' was not such an urgent word as it sounded, and I might have waited a few days as the ship is nowhere being ready for sea. S.S. *Mantua* is one of the modern P. and O. liners, and the day war was declared she had just returned from a cruise in the Baltic with a shipload of tourists. The ship is 11,000 tons

* His cousin, Admiral Sir Herbert Richmond.

and is now being converted from a floating hotel into an armed merchant cruiser—presumably to patrol trade routes or something of that kind. She is being armed with eight 4.7 guns; four for'ard and four aft. My boss, Fleet-Surgeon Jackson, is an easy-going Irishman and I think we shall get along all right. Captain Tibbetts, M.V.O., R.N., in command.

The process of conversion is an exceedingly noisy and dirty one. Hundreds of men swarm about the ship day and night, camping in the cabins and what were the saloons, eating meals at all hours and generally appearing to do nothing. Yet somehow things are getting done, and in these few days there are decided visible changes—such as strengthening the decks in places for the heavy guns we are to carry, etc. I wonder when we shall fire them? Altogether, it is a very noisy, dirty, uncomfortable time, and what a waste of these glorious days.

There is another big liner being converted in this dock, and to her have been appointed two very young Temp. Surgeons—volunteer medical students from Bart's. Both agreeable fellows but not overburdened with brains. The question of who was to be the senior of the two was settled by the fact that one of them had once done an amputation and the other had not.

AUGUST 14. (*Tilbury.*) I got into uniform yesterday for the first time, and felt very conscious of eight flaming brass buttons and two stripes of gold braid with a strip of red (for blood presumably) between them, which are the recognition marks of my rank and quality. The peaked cap is an abominable thing.

This evening I walked over to the railway station to buy papers, and was mistaken by a lady for a railway official. She was much more embarrassed than I was when she found out her mistake, and it showed me that I am not nearly so conspicuous as I seemed to myself to be.

AUGUST 15. Went up to London to see Ogilvie-Grant about the safe bestowal of all my New Guinea things: my precious stone axes and knives and so on, which I had hoped to dispose of in America. O.-G. laughed at me for saying that neither of us will live to see the end of this war. He says that it cannot possibly last longer than a few months and that then we shall have a real peace without the former German competition. As for its being finished in a few months, who can really believe that if he remembers that it took us thirty months to come to terms with the Boers? Thirty years is doubtless too long a guess, but I don't think it is so wide of the mark as Christmas is for the end of this. Heaven send I am wrong.

A merry party of riveters have just started within a few yards of my cabin, and they make a terrific noise at night in order to earn their higher wages, so no more of this.

AUGUST 17. This is not, as I thought at first, a transport business, but we have a roving commission, and are supposed to patrol trade routes as 'commerce destroyer' and 'commerce protector'. By all accounts there do not seem to be many German merchant ships left at sea to destroy, nor will our own ships require much protection when the few German cruisers have been accounted for. We are armed, but I do not suppose our guns will ever be fired in anger. Nobody knows where we are going; possibly the Mediterranean or along the South American route. We are full of coal and I expect it will be long before we see land.

AUGUST 20. Out of dock and into the river. It began to be whispered that we were going north, and when we were off Dover, Captain Tibbetts told me that we go first of all to the Orkney Islands. This is a great blow, for we have been all the time led to believe that we were going south along one of the trade routes as 'commerce destroyer' and 'commerce protector'. We have all got white clothes and I have brought no warm things at all, so I shall be in a desperate plight.

Kept close in shore as far as Dover where we dropped our pilot, and then out to mid-channel; fine weather. Passed the Scillies shimmering in a last haze, then up through Irish Sea. We number nearly 400. The seamen are for the most part of the R.N.R., fishermen from the east coast—Grimsby, Lowestoft and Yarmouth. As fine a lot of men as you would see anywhere. The training most of them have had amounts to very little, but they are learning their gunnery and drill quickly enough. The stokers are of a very different kind, being of the less desirable class of East-ender with a strong dash of Liverpool Irish. They sign on under Board of Trade regulations, but are subject to naval discipline and they are constantly giving trouble. Besides these is a detachment of forty Marines, nearly all of them recruits and good steady fellows. The only objection I have to them is in their drummer, who delights in sounding his bugle unexpectedly behind your back. I have managed to keep him on the sick list for the last week, so we have been mercifully free from his horrid blasts.

AUGUST 23. Rathlin Island....Lay up near Tory Island. Fired four practice shots from each gun; tremendous din but no damage to ship...then off north. Rather thick and saw nothing till Hebrides....Steamed into Scapa Flow, an inland sea among the Orkneys. There found about fifty colliers and several men-of-war of different kinds; soon afterwards eleven battleships. The 2nd Battle Fleet entered Lerwick Harbour to coal; looked very fine.

AUGUST 25. Our ship has been inspected by Rear-Admiral De Chair (H.M.S. *Crescent*), commanding the squadron in this region. We all lined up in our best toggery, and after that we got our orders, which are to patrol the sea between Shetland and Norway a little south of Bergen (along 60th parallel and about 120 miles straight backwards and forwards), and to stop and search every blessed ship we come across. So we have really been given a very honourable post at the mouth of the North Sea and on a line that will very likely be taken by any German ship that gets out. We should be able to put up a very fair show against any smaller ship than ourselves, but heaven help H.M.S. *Mantua* and all in her if we run up against anybody with a 6-inch gun or bigger. We have absolutely no protection. We are an enormous target (far bigger than any battleship), and we should go to the bottom like a stone.

AUGUST 26. Over to Norway and back; very thick weather. News by wireless this evening that German submarines have been seen east of the Pentland Firth, about 80 miles south of us; hope they come no further north.... Sounds of distant firing about 7 p.m. but no information.

AUGUST 27. No news, as currents were interrupted by enemy....Norway and back as before. Have stopped and searched vessels near Norwegian coast.

We have constructed what is called the 'wardroom' by screening off with pieces of canvas a corner of the former passengers' saloon. It is a bleak and cheerless place and it will be dreary enough in the winter. Talking of winter reminds me that as a precaution against the cold weather several of us, including myself, have started growing beards, and a proper lot of pirates we look in our various stages of unshavenness, particularly the Skipper.

AUGUST 29. Back to Lerwick where we got different orders, and are now to steam up and down between Orkney and Shetland. We pretend not to bother about submarines, etc., but they do get on some people's nerves.

The weather has been glorious: sunny days and moonlight nights, with generally land in sight or whales or porpoises or birds to look at. We patrol up and down on our beat at a leisurely ten knots or so, turning round every few hours, and varying the exact beat according to instructions received by wireless. Every ship seen is challenged, and if her reply is not satisfactory we send a boat off with an officer to board her. Astonishing quantity of shipping through these channels; crowds of passengers from America, and one feels pretty certain that numbers of them are Germans coming back to fight against us, but as they are in the shelter of a neutral flag we cannot touch them. So far, it cannot be called a very close blockade, as there is at present only one other ship—the *Oceanic*—on this patrol, and there is consequently plenty of room for ships to slip through, especially at night. We steam without lights at night and all the scuttles are blackened and tightly shut. To strike a match on deck after dark would be more than your life is worth.

Have seen some Great Skuas. They look immense birds on the wing, and have a curious heavy, almost eagle-like flight. When a Great Skua chases a gull he amazes you by the quickness of his twists and turns. I have now seen a great number of them, particularly round the west side of Foula Island on one of the sheerest and highest cliffs I have ever seen. One day we steamed close into the famous cliff where the last British sea eagles nested. I thought I could distinguish the eyrie, but not a sign of the bird. The last I heard of them was that only the hen bird was left and she was almost white with old age.

Up through the Minch, whence a splendid view of the Coolins in Skye. I could clearly make out many of the pinnacles I have climbed. Handa Island was easily recognizable with its white splashing of birds, and Ben Stack towering up behind it. What good times I have had up there!

Up and down, stopping steamers of all sorts, trawlers, tramps, and to-day a small Danish mail-boat from Faroes to Copenhagen. On board was an English lady passenger who did not want to be taken through mines to Copenhagen so asked to be taken on board of us. She was informed that she might not be landed for fourteen days, but still she elected to come. Amusing to see how Captain and others hang about a skirt; one among more than 300 men. She is a fine strapping athletic girl—secretary to a dentist in Harley Street—hospital nurse type.

Thank heaven for wireless! We get the 'Press' every night; quite long messages sometimes, but at the best it is only enough to make us want more. Such tremendous things are happening, and we get only a meagre sketch of them. The ships we intercept occasionally have newspapers but more often than not they expect to get news from us.

AUGUST 31 *and onwards*. Still stopping everything that sails or steams. The trawlers are always suspicionable, as the Germans have captured a good many of ours and are sending them to sea laden with mines.

We hear from an Aberdeen trawler that thousands of Russians from Arkhangel have been landed at Aberdeen and other Scotch ports.

SEPTEMBER 10. Calf of Man....Bar of River Mersey.... L'pool so full of ships that we are kept waiting outside until room can be made. Into river about 5 p.m. and then into Canada dock. Go ashore for lunch and business. Beastly place L'pool: coaling filthy. Dined at Adelphi Hotel and went to a music-hall—all very depressing.

SEPTEMBER 11–14. Coaling all the time; about the record for bad coaling—four and a half days for 1900 tons. Found it very difficult to get out of dock as it was blowing such a gale; bumped and strafed and did ourselves a good deal of damage.

SEPTEMBER 16. Cape Wrath....Fine view of familiar Sutherlandshire mountains. Soon afterwards a big ship was seen coming down at a great pace from the north, and as she would not answer our distant signals we altered our course and began to run as fast as we could towards Scapa Flow, supposing it to be an enemy's man-of-war. Bugle sounded 'action' and for a few minutes great excitement, until they condescended to tell us they were H.M.S. *Africa*.

Up along west coast of Shetland, Mainland and Unst. Fine cliffs on Yell where the sea eagle used to breed.

SEPTEMBER 18. Across to the Norwegian coast, patrolling up and down between Bergen and the Naze. With us—7 miles on our beam—is the *Alsatian*; same kind as ourselves. It seems that we are here as a sort of lure to entice out any enemy ships that may be—and probably are—lurking in the Fjords. A very unpleasant rôle to play, as we have neither speed nor guns worth anything. Everybody very sick about it, from Skipper downwards. Cruisers suddenly appear on the horizon, come round us, and shoot off again. Evidently something going on but we don't know what.

SEPTEMBER 22. We were on our way to Balta Sound when a signal came from the *Iron Duke* telling us to go over to Norway and try to intercept a German collier preparing to leave Trondhjem. The confounded captain of the *Teutonic* talks all day and night to us by wireless and gives the whole show away!

SEPTEMBER 24. Up and down off entrances to Trondhjem Fjord. *Teutonic* wires that she has entered the Trondhjem Fjord and invites us to come with her and seize German collier! In Norwegian waters!

SEPTEMBER 25. Signal from *Iron Duke* saying we are not allowed to infringe neutral waters! Furious gale, very big sea, ship all over the place; hove to all day.

SEPTEMBER 28. Signal from *Iron Duke* telling us to keep a good look-out for German submarines as the enemy is probably aware of our position. Have no doubt they are. Not very great sport for us, but I hope that the sea is too rough for action by submarines. Sure to be some lurking in Norwegian waters, territorial or not.

OCTOBER 1. Back towards Shetlands, then to L'pool.

OCTOBER 3. (*Liverpool. To his Mother.*) I meant to write to you yesterday, but there was such a mess with coaling that it was not possible. To-day there is a fearful hammering of iron by the repairers, so it is equally difficult. In spite of the dirt and din it is rather a comfort to be here after the buffeting we have had this last fortnight. Fearful gales—very disagreeable. The days are beginning to get very short now in the latitude of North Scotland, and in a month or so there will only be a few hours of daylight, but we are hoping we shall not be kept up north all the winter. I should greatly prefer to go into the Tropics or thereabouts. My shipmates are a very decent lot and we get on all right together, but I sometimes wish for some more understanding persons to talk to.

We cruise up and down, backwards and forwards, night and day, searching again and again, and so on and so on, and that is the long of it and the short. Occasionally we have alarms—actual or artificial—which serve to keep us awake and drilled to do the right thing at the time of emergency. How tired of it we shall be before there is peace. Perhaps things are moving a bit now, but I am afraid it is only the beginning of a very long story.

Furious banging within 6 feet of me and I cannot write coherently....

OCTOBER 10. North again. Fearful alarm over a submarine seen by the *Teutonic* 40 miles away from us and reported by wireless. We came to the same place a few hours later and several shots were fired at the supposed submarine. We all rushed to quarters, prepared for anything that might happen; nothing did. Am not greatly struck by the manners of the R.N.R. officers who jeer at all things R.N. after it is all over....

OCTOBER 12–29. Ordered to Scapa Flow. There found the 1st, 2nd and 3rd Battle Squadrons coaling. Skipper went on board the *Iron Duke*, lunched with Jellicoe, and came back with orders that we are to proceed at once to Arkhangel for a special purpose—not to be divulged. On our way out I noticed that two of the entrances to Scapa Flow had been blocked by sunken steamers. We are informed that German submarines have been all over the place and that one had two shots at the *Iron Duke*—so I have no doubt that what we saw the other day *was* a submarine, and we were lucky to escape.

The first land sighted was the North Cape, and after skirting along the coast we got to the Murman coast; low and very black, flattening out as you approach the White Sea. As the ship is too big to go up to Arkhangel itself we anchored off the mouth of the Dvina River, some 25 miles from A. We learnt the object of our voyage when, on the following day, a steamer came down the river and proceeded to discharge into us (and the *Drake* also) about eight million solid golden pounds, in boxes. The golden steamer was guarded by an escort of 100 or more Russian soldiers looking very fierce with beards and furs. Some officials came on board of us and bewailed the liquor restrictions in their country, all the time making the most of their present opportunity. I was assured by one of them that not a single Russian soldier has left Arkhangel, or any other port, for England or France. So much for the story of the snow and fur left in British railway carriages! We spent four days at the mouth of this river, and they were still cloudless days, freezing in the shade at midday but very warm in the sun. The nights were brilliant, with gorgeous displays of A.B., and clamorous with the cries of wild fowl. I never saw anywhere so many geese and duck. String after string of them flying over in every direction, but so far as I could see not definitely migrating. Altogether it was a time that made one glad to be alive, fit to walk across the world instead of being cooped up in a floating box such as this is!

It has been decided that as much woodwork as possible shall be removed from these ships, so we have recently submitted to having our insides torn out by a gang of men armed with picks and axes. Now I have a weakness for good carpentry and a pretty piece of wood, so it was sad to see the beautiful fittings hacked out and piled in heaps of fragments along the dock side; but we were in a hurry and could not wait.

Sailed North Cape—about 100 miles to north of it, 73° 45′—then due west for 500 miles. Saw two glaucous gulls, and some smews of an almost silvery whiteness in their winter plumage. The latter were so tame that they let the ship pass them within not more than 20 yards.... Passed outside the Hebrides, and even outside St Kilda and Flannan.... When we were within a few hours of steaming up the Mersey we received a wireless to look out for submarines. At once to 'action', and all eyes skinned for periscopes. Great tension all day. We, and the *Drake* also, started zigzagging—the new plan for dodging submarines—and kept it up until we reached the river. A few hours later we learned it was not a question of submarines but of mines laid north of Ireland; H.M.S. *Audacious* had hit one of these and had been ultimately lost. So we had come through this minefield in such a way as to double our chances of being blown up, but luck was with us! I cannot help wondering whether the people who laid the mines had not an inkling of our approach, for had they succeeded in bagging either the *Drake* or the *Mantua* they would have sunk several millions of golden pounds. It was a great relief to get rid of our precious freight. Spent almost a week at L'pool, where a blank dock wall partly obscured by drizzling rain was not altogether an unpleasant sight to see through the scuttle instead of the eternal sea. I had hardly been out of the ship since I joined on August 8. It seemed years ago. We had certainly earned a short spell ashore, yet one felt strangely uncomfortable walking about a town in days when such tremendous things were happening, and I for one was not sorry when we pushed off again, though I knew the prospect for the next twenty-eight days was not a very cheerful one.

NOVEMBER 7–20. Patrolling up and down between Faroe Islands and Iceland. The North Sea is now closed and we are the last part of the fence. Gale, gale, gale; only two ships seen all the time—a trawler and a small sailing ship. Met the *Drake* one day and exchanged old newspapers. Intervals now and again for coaling, and at Birkenhead I had a few days ashore. It is an immense relief to meet and talk with one's friends after months spent in the company of one's messmates.

DECEMBER 20. (*To Mrs J. H. Clapham.*) This is a desperately stagnating existence and one misses the small varieties of shore life. Myself, I would far rather be alone with Dayaks in New Guinea than cooped up with 350 white men on board ship. Which reminds me that William and his friends have another crime to their account. I had more than half-made plans for another expedition to New Guinea in 1915–16. Now that will never come off, even if the war ends next year or the year after, as nobody will have a penny left to spend. But apart from selfish considerations it is all a very desperate business and I don't suppose that we of our generation will live to see normal times.

Seeing the sad state of some others, I have been glad these last months that I have no wife and no children. I am sorry this is all rather lugubrious, but one must grumble about something; it is a privilege of those who go down to the sea in ships, and is a sign of health and well-being.

I had a terrible dream of bombs or shells being dropped

on Cambridge—most vivid. I was high up, in (I think) Gibbs' Buildings!...

JANUARY 5, 1915. There are temperate intervals between the gales, and one of these occurred by the mercy of heaven on Christmas Day, which we celebrated in real sea fashion. After 'church', the padre and all the officers, headed by the captain, proceeded round the ship with a 'funny' party of half a dozen individuals dressed in outlandish clothing and making din indescribable on drums, trumpets, whistles and so on. At each mess we stopped and tasted (or appeared to taste) the plum duff of the mess, and the captain made a speech. Wonderful people these sailors! They had decorated their messes with garlands, flowers of paper, inscriptions of welcome, and all with as much care and ingenuity as if they had been at home. In the evening we had a sing-song, and it pleased my democratic heart to see officers and stokers swinging hand in hand to the tune of A.L.S. If the truth must be told, the wardroom was not more sober than the lower deck, for we were all glad enough of an excuse to forget the turmoil and the buffetings of the last two weeks. But that same night began one of the worst gales I have ever imagined. A sea of colossal size—so great that you would not believe it possible—and I have no hesitation in confessing that I was horribly afraid. We were 'hove-to'—that is heading as nearly as possible straight into the wind—with only enough speed to give us steerage way—and at the end of the third day we were so far out of our proper patrol that it was necessary to turn round. Now the first part of the turn is comparatively easy, that is until the ship is broadside on to the wind: then comes the danger. The wind, blowing with terrific force against the whole bulk of the ship's side, does its best to hold the balance between the bow which wants to go down wind and the stern which wants to come up; the only thing that makes a difference between the two being a little bit of a rudder, which is of no avail unless the ship is moving fast enough to get steerage way on her. Then comes another difficulty. If the engines move fast enough to get speed on the ship, there is desperate risk—when the propellers come out of the water and 'ease'—of doing damage to both engines and ship; thus you have the alternative of too little speed and not being able to turn round, or too much speed and knocking the engines to pieces. Meanwhile the ship lies broadside on to the gale and one after another enormous seas roll up and seem like to swamp her. I can't describe these seas. They are not waves but moving mountains of waters. In the sudden wonderful lull in the wind which comes as they approach, you wonder for a moment how the poor thing can rise to such mountains. And then, Heaven help us when we near the top, and in a flurry of spume and spray the wind lifts sheets and tons of water and hurls them down upon us! Here again, wind is not the word for the formidable force that lives in these latitudes. A ship of this kind is built to carry several thousand tons of cargo, which gives her a draught of about 30 feet in ordinary conditions. As we are now, we are an empty ship in ballast, with enough weight in her to give her stability but with considerably less draught. I think some of my most miserable moments have been when the ship turns round and rolls broadside on to the seas. As she goes over and over you instinctively hold on to something—though what good that does I don't know—and then for one horrible moment she pauses and swings back again. It is all so wearisome and one's body cries out for rest. If you try to read or write you must have your body in a more or less secure position, and at night-time in the half sleep that comes to you, you instinctively do likewise. Exercise is of course out of the question, and at the same time it is unnecessary, for the constant balancing is continuous exertion.

Wonderful little craft these trawlers, and they ride over the huge seas as lightly as seagulls. During one of our worst days, we passed a Yarmouth boat lying hove-to, her decks quite dry and nine of her crew sitting contentedly smoking on deck in the lee of the deckhouse. We were plunging about—pouring wet from stem to stern. There are usually two or three trawlers on the fishing grounds between the Faroes and Iceland, and many go round to the White Sea banks. The men have brothers and cousins on board our ship and I am told that these little boats can make as much as three or four thousand pounds in a six weeks' voyage. Many of them are lost every winter, and I don't think anyone would grudge the price of fish who had seen the winter conditions up here. These east coast fishermen are as fine a crowd of men as you could see anywhere and certainly the finest fellows on this ship. I think they appear at their best when boarding ships. In bad weather this boarding of vessels is a difficult and dangerous operation. To lower the small seaboat from the davits forty or more feet above the water, then cast off both shackles exactly at the right moment as the wave surges up from below,—that is a fine test of seamanship, for a single hitch would mean upsetting the boat and the almost certain destruction of the crew. They sometimes have a heavy pull of perhaps half a mile through overwhelming seas to the other ship, which as often as not lies rolling heavily in the trough instead of turning to make a lee for the approaching boat. The inspecting officer must await his chance, jump at the swinging ladder, climb up the ship's side and get through with his job as best he can; and when the whole reverse process has been gone through I think we all heave a sigh of relief when our boat is hoisted in again. It often happens that a ship's papers are not satisfactory—I doubt if any of them really are,—and in that case she is sent to the most convenient British port in charge of an officer and an armed guard of marines, who are glad enough to get away for a few days from the ship's routine and the deadly monotony of the patrol. When the weather is so bad that boarding is impossible we order the ship to follow us, and pass her on to the patrol south of us who in turn shepherds her to less tempestuous regions where she can be examined. I wonder how many ships manage to slip through the patrol? A certain number must do so, as our beat may be 60 miles long or more, and the range of vision is very short—especially in these dark winter days. A neutral or enemy vessel bound from America and really wanting to pass through the patrol should stand a fair chance of doing so by waiting 100 miles to the west, taking careful note of our wireless, and then trying to pass through in the dark. It seems to an outsider like me, that our ships make a quite unnecessary use of wireless. We are constantly signalling our position and course, giving or receiving orders to alter the patrol, and making various other signals all of which are heard by anyone who cares to listen. Of course they are all made in code, but the insoluble code has yet to be made.

JANUARY 6. I find these long hours of darkness very trying. It is not really broad daylight until about half-past nine and it is black night again before four o'clock. Walking on deck is almost impossible, and below decks there is no daylight, as all scuttles, etc., were blackened months ago. Twenty-three hours a day of electric light is bad for the eyes and I begin to feel it. Happily I have a very good friend in the captain who often invites me to his cabin on the bridge, where we talk of most things under heaven. He is the only other I have found in this ship who believes that we are in for a long war; the others all thought at first that it would be over by Christmas, and now they are convinced that it cannot last a year. We shall not be beaten, that is certain; but can we win?

JANUARY 24. Hear that H.M.S. *Viknor* (armed merchant cruiser) has been lost during the week somewhere off the Hebrides. Probably rolled over in one of these infernal gales. Later, heard that the *Clan Macnaghten*, another of H.M. armed merchantmen, had gone astray also near the

Hebrides.... The after compartment of this ship is full of water to above main deck and cannot be pumped out. Water-tight doors and bulkheads hold well, so nobody seems to think much of it.

FEBRUARY 12. We are now between the Faroes and Iceland, and have been out of sight of land for twenty-two days. I always imagined the Faroe Islands to be flat, but they are actually very Norwegian in appearance—big mountains—now covered to the sea with snow and intersected by deep Fjords. It is said they are the lurking place of German submarines, but we still respect the 3-mile limit and do not hunt them out. I confess I cannot see the reasonableness of that, for when we are at war with an enemy who breaks the rules we should meet him on his own grounds in self-preservation, even at the risk of offending the Danes.

St Kilda more than comes up to my expectations. Outlined against the sunset it is almost Alpine, and the cliffs are magnificent. Even at this time of the year there are immense numbers of sea birds about the islands, mostly fulmar and gannets. I should like to know more about the life history of the fulmar. During the long days of last August there were some following the ship at all hours of the day and night—indeed they seemed never to go to rest at all,—and even now, when we are very far from land, I can see them so long as it is possible to see anything at all. In the heaviest of weather we are seldom without five or six of these birds. It has been very interesting to me to see such bird places as the Flannans, Sule, Skerry, Sulaskeir, North Rona and St Kilda. In fact I believe we have seen all the outlying British rocks except Reskall, which is a long way west of our patrol. We have also seen two historical spots which Newton and H.B. used to visit: the Shiants, where the sea eagle used to breed, and Papa Westray, the home of the last British great auk. This reminds me that my life of Newton* ought to have seen

* His *Life of Alfred Newton* was eventually published in 1921.

the light about this time. I had reckoned on having it finished before Christmas, but now goodness knows how many years it will be before I get back to work on it.

To go back to the blockade and ourselves, we are now officially styled the 10th C.S., and we number more than twenty ships, that is more than any other squadron ever heard of. We should make a strange show if we were all to meet—banana boats, western ocean liners, ditchers, big cargo boats, some one and some two-funnelled and at least one with three, and with such comic names as *Columbella* and *Changuinola*. Though we are in touch by wireless with the next ship in the patrol we rarely see any of them except at a great distance, but one night not long ago we nearly saw a great deal too much of one of the squadron, and it was only by the very prompt action of the officer on the bridge that we did not collide. After a long spell of heavy weather, currents are very strong, and it is easy to be miles out of your proper position,—especially if it has been impossible to take a sight by sun or stars for several days. This, together with the fact that we steam without showing a glimmer of light on the blackest nights, makes the possibility of an accident not uncommon. The *Clan Macnaghten* disappeared utterly without leaving a trace behind her. A curious thing happened in connection with her loss. We went to search over the area in which she had last been patrolling, and in broad daylight I saw through my glasses at a distance of less than a mile from the ship, a boatful of men. I saw it most distinctly for a few seconds and then lost sight of it, for we were rolling heavily in the big sea that was running. I raced up to the bridge and reported what I had seen, but nobody else had seen anything, and we saw no more sign of it. I suppose some would call it a vision or a hallucination, but even so it was extraordinarily vivid to me.

We understand very well the importance of this blockade business and I don't think we grumble over much, but we do get a little annoyed when we hear what a large proportion of the ships sent in for examination are released to carry their cargoes of cotton and oil and foodstuff to Sweden, and so to Germany. No doubt the F.O. has more troubles than we wot of, but we cannot help thinking that they are prolonging the war by these methods of peace. There is a big trade in ponies between Iceland and Denmark, whence the poor little creatures go to Germany presumably for transport purposes and eventually for the sausage machine. For a short time this was prohibited and we captured the pony ships; now they are free again; heaven knows why. Not long ago one of our officers boarded a pony steamer in very rough weather; the rails and pens in the 'tween decks had mostly been smashed, and at every roll of the ship the ponies (there were 192 of them) slid from one side of the ship to the other in a squealing kicking mass. I remembered Conrad's description in *Typhoon* of the Chinese coolies in the hold of a ship.

Sailing ships are very rare these days, but not long ago we went close alongside a small brigantine homeward bound from Reyk to Bideford, and it was pleasant to hear good West Country voices again after the semi-Cockney tongue of our ship people.

There is one thing we do get in plenty, and that is fresh air. The regular man-of-war is constructed so that every part of the ship can be made as far as possible air-tight and water-tight, but our *Mantua* was built for coolness in tropic seas, and the designers were at pains to let the greatest amount of air into the ship—much to our recent discomfort. It has had some good results, such as this: the last time we went down to coal, I was reading in my cabin with the scuttle open—it was a rare calm afternoon with not a vestige of a breeze—when suddenly I became aware of the smell of peat smoke—quite unmistakable. I went on deck and found the north of Ireland right ahead of us,—7½ miles, according to the navigator—to the nearest land. Pretty good smelling that; only possible after thirty days of north Atlantic gales!

In spite of discomfort this life is a healthy one, and the sickness among the ship's company is almost negligible. My medical work consists almost entirely of cuts and bruises, which could easily be treated by a medical student; in fact there is generally more medical work to do in the four or five days in harbour 'coaling' than during one month at sea. Liverpool is the place we usually go to for coal, and personally I am always quite glad to see the last of it, even though the prospect of a month of storm is not very enlivening. What I feel when I am ashore is, that one has no right to be loafing about a big town when the Army is up to the neck in mud in France. We at all events have our dry beds and hammocks to sleep in, and plentiful and regular food; we have no business to grumble. In Africa and in New Guinea we were often a great deal worse fed, more uncomfortable, and in not nearly so good a cause as this. It looks very much as if my third expedition to Dutch New Guinea would never come off. I had got promises of the necessary funds last summer, and the F.O. was in process of obtaining the sanction of the Dutch Government; moreover I had agreed with two splendid fellows to accompany me, and we were to have left England this May. Now, who knows?

FEBRUARY 19. (*Glasgow.*) Have got forty-eight hours' leave, but I don't know yet where I shall go to. It will probably be to the railway station where I shall ask for a ticket to a quiet place far from ships. The last month has been rather a poisonous time. Happily the days are getting a bit longer now, and I hope that by the end of next month the weather may begin to moderate a bit....

FEBRUARY 23. Went to Trossachs Hotel. Most beautiful place in winter with nobody there. Unluckily I was flattened out by acute lumbago and could hardly get back to ship, where I was on sick list for seventeen days.

MARCH 1. I have a superstition, or perhaps it should be called a self-persuasion, that spring begins on St David's Day, and I usually celebrate it by making an excursion of some sort. At Cambridge it used to take the form of a walk in the Fens, sometimes in sunshine and sometimes in snowstorm. But of many firsts of March, I can remember none so cold and bleak as this in 61° N. lat. It was chilly enough in the dark days of December and January, but with the lengthening twilight the cold has become far more intense and it is now infinitely more bitter than in mid-winter.

MARCH 25. (*To J. H. Clapham.*)...It is splendid news that U 29 was cut in half by the *Temeraire* last week. The Blermans (that is a portmanteau word of my own) cannot have an unlimited supply of these subaqueous craft, and taking it all round they have done precious little damage considering the chances we give them....It is difficult to see King's with only thirty men in residence, and the New Buildings full of nurses. I wish I could come up and have a look at you, but that is not possible now. Instead, I will drink all of your healths to-night, and I have no doubt that you will be saluting us. Before this—when I have gone off on somewhat desperate ventures—and I never told you the half of the risks that there were in New Guinea—I have been quite sure I should come back again to dine with you in Hall; but this time I have a very vague uncertainty about it, and there is a very strong probability that we shall not meet again in Cambridge. Very likely I am wrong, but I have a curious kind of instinct in these things. I don't seem to have anything to tell you, though if we met I have no doubt there would be plenty to talk about.

MARCH 31. Got a few days leave, and after going home went to Kew and walked through the Gardens. The early daffodils are quite magnificent, but a great many things have been cut by these hard nights.

APRIL 12. (*To M.M.*) ...It is a most glorious spring day, and I just tingle with a longing to be on land and to smell the good earth. I cannot think why the Almighty made this huge desert of waters. It seems such awful waste of time to be ploughing up and down at sea, and I grudge every moment of it. Like you, I should like to go to sleep for a few years and wake up when it is all over. The worst is that at my time of life (on the very verge of forty), one cannot afford to waste even a few months, much less a few years. There is such a tremendous lot that one wants to do and so short a time to do it in. I wish I could believe in reincarnation, or indeed in any other comfortable belief. I am sorry to inflict all this grumbling stuff on you but much solitude at sea is bad for one's good manners. I do so much admire your pluck in wandering alone about the Caucasus. I should like to hear a lot more about it. It has always been one of my most cherished dreams to go there, but again, this dreadful business has knocked all that sort of thing on the head.

I am deeply engrossed in *The Pastor's Wife*, which I expect you have read. It is a very chastening book for a man to read and I have learned much. Happily the sadness of it is relieved by a most engaging humour and I often startle myself by laughing aloud. There is plenty of time for reading here at sea, but it is extraordinarily difficult to concentrate on anything in a wallowing ship, and I have done nothing solid these eight mortal months. What a life!...

APRIL 25. About midnight last night—fine and clear but nasty in patches—we were coming down from our patrol and rounding a headland, when a small error was made which brought us on the wrong side of a light—consequently on to an outlying piece of Scotland. We were all on deck in a couple of minutes and at our proper stations, but it was soon clear that there was no immediate danger provided that we were not detected by lurking submarines. Happily it was nearly low water and we were able to get off after a few hours, or it might have gone hard with us. It is a nasty thing, that sound or feeling of running on to rocks. Running ashore, even when it is done so gently as we did that night, is quite a different thing: you feel at once that the ship is dead. After a few days at sea you become so accustomed to the hum of engines that you wake up from your deepest sleep when they stop, but as the motion of the ship continues you turn over to sleep again.

APRIL 26. Dry dock at Govan. Find that the ship's bottom is badly damaged and that repairs will take at least a month. Govan is a beastly place and the dry dock is the beastliest place in Govan. Baths cannot be used on board and sanitary arrangements are a quarter of a mile away; add to this brilliant sunshine, clouds of dust and the constant hammering day and night on the wounded ship, and anyone will agree that it is not a pleasant life. The streets of the city are thronged at all hours with crowds that wear a holiday rather than a wartime air. Certainly there are plenty of soldiers about, but there is an enormous majority of young men in plain clothes. One is reminded very much of the streets of London in the recruiting days of the Boer War, when 'Tommy T.A.' was at the height of its vogue. The amount of drunkenness is perfectly appalling. I have seen more people, both men and women, intoxicated in G. than in any other place in the world in so short a time. The effect on the ship's company is very bad and I shall be glad enough when we get away again. I wonder if we shall resort to conscription in the end.

MAY 6. Went for a week to the Loch Awe Hotel and had three or four days' fishing on the lake. Wonderfully beautiful place in early May, and you can imagine how it appeared to me who had seen nothing for months but the North Atlantic and Govan dry dock! The boatman recommended me a bay a few miles down the lake, and suggested we should troll for one of the big fish on the way down. So we went to the post-office shop and purchased a villainous-looking red phantom minnow, which the girl said looked 'tasty'. I had only a light trout line, a single sea trout gut cast, and my rod was a little 10 foot whole cane which I bought twelve years ago. We had not pulled a quarter of a mile down the lake, when there came a tremendous tug on the line and the reel began to shriek. The boatman knew his job and followed the fish, and fortunately I had 40 yards of backing on my 60 yards of line. Somebody years ago advised me to do that in case of eventualities, otherwise this fish would have been away at the first rush. Then followed a most exciting time. With such tackle as I had it was impossible to put any weight or strain on the fish, and he towed us all over the place just as he liked. His first most dangerous rush was up to the head of the lake where the Orchy comes in and the weeds are, but we managed to steer him clear of those. Once he jumped, but directly between us and the sun so that we could not see his size, and then he must have run in quite close to the boat as the line was slack for a long while. I reeled wildly, but he was still on. Later he went close to the shore, and I jumped on to the beach hoping to pull him up and tail him, but he was off again like a torpedo, and we tumbled into the boat and took up the chase. After some time he showed his back once or twice, and it looked as if he were tiring. We shouted to an old man who was watching us from the beach and told him to bring us a gaff. After about twenty minutes he came off in a boat bringing a thing like a shepherd's crook, which he passed over to us. It was some time yet before I managed to reel in and get close to the fish as he was rolling on the surface, but the boatman gaffed him cleverly with his clumsy weapon and landed him in the boat where the gut parted at his first kick. One more rush and jump out of water would have saved him, and I should have struggled for an hour and thirty-five minutes in vain! We carried him up to the hotel and weighed him carefully—21 lb. 14 oz.; a clean fish straight from sea. How is that for my first salmon?

I spent a day motoring with an officer on 'sick leave' up through Glen Cor—where snow still lay deep in the gullies—across a howling waste to Tyndrum, Loch Lomond, and Inverary; back to Loch Awe. We went through a hundred miles of the most beautiful part of the Highlands, and so far as we could see the whole world was at peace. I don't think either of us gave a thought to the war until we got back to our hotel and heard the first rumour that the *Lusitania* had been sunk.

What unspeakable devils they are. But after their diabolical chlorine gas attacks can one really be surprised at anything they do now? When they advertised in New York that they were going to sink the *L.* I was so convinced that they meant to do it that I offered odds of ten to one on their making the attempt. (No, I did not collect the bet; that would be a bit too much like blood money.) But I am afraid the Navy will be greatly blamed for this—and, as it seems to me, rightly. Of course they cannot spare cruisers or destroyers to escort every transport that comes across the Atlantic, but one can't help thinking that they ought to have made an exception in this case after being so openly warned; or do the authorities still believe that the enemy is honourable and a gentleman?

MAY 10–26. These days have been quite perfect: bright sun and a cool wind: trees getting green: swallows and cuckoos and all the other signs of spring that I have not yet seen. Left Scotland and went home. Spent an afternoon in London at Kew where I found the bluebells at their best. On my return to Glasgow I found an order from the Admiralty telling me to join the *Agincourt*, 'forthwith'. (That mysterious Navy word.) As the *Mantua* was going to sea in two days' time I said good-bye to my three good friends with real regret. It is curious how one gets attached to a ship, even to a floating barn like the *Mantua*, but she had been my home for nine months—which is a longer time than any house has been these many years.

I went by train to Inverness and then to Thurso. At the wayside stations, instead of the usual sportsmen with guns and rods and gillies and stalkers, there were small parties of Royal Scots going or returning from leave. I did see one lucky fellow playing a fish in the Helmsdale as we went by, but none other anywhere—and that in the last week of May. At Thurso I was met by an officer who told me what to do with baggage, where to go for tea and how much to pay for it, how to find the ramshackle car that would take me the 2 miles down to the steamer, and so on. It made me feel just like a small boy going to school. But the Navy nurses its children well and saves a deal of trouble. Although the P. Firth was like a boiling kettle and the little Hebridean tumbled about like a cock, my training of the last nine months saved me from the fate of some of my fellow-passengers. On the other side we were sorted out into drifters and dispatched to our different squadrons, which got under weigh and disappeared in a cloud of smoke as we approached them. Personally I was not at all sorry, as I did not like the prospect of being dumped into a strange ship at nearly midnight. I spent two days in a kind of depôt ship until the fleet returned from sea, and then I was sent on board the *Iron Duke* where I found Hilton Young masquerading as officer of the watch with a telescope under his arm—neither of us knowing that the other was in this service. It was very cheering to see a friendly face in all the multitude of strangers. Then came the awful plunge into the *Agincourt*. I joined her in fear and trembling. She is the largest ship in the fleet, fourteen 12-inch and twenty 6-inch guns, 1250 complement. However, like most things of the kind, it was not actually nearly so alarming as it had been in prospect. For one thing they were coaling when I came on board, so it was easy to escape notice in the dirt and noise, and later on—when they had cleaned themselves—I managed to meet the other officers at intervals in ones and twos, which was a great deal better than being suddenly pitched into the midst of a whole crowd of them; they are not so very fierce after all!

My old fishing rod has done me another good turn. It was reported to the Commander,—who is a very keen fisherman,—that I had come on board armed with a rod, so he invited me to go away fishing for a day or two soon after I joined the ship. We went to a small lake on the other side of the harbour and caught some nice little trout; afterwards to a farmhouse for tea, with fresh scones, eggs, butter and honey—like the teas of old days; and so back to the ship in late dusk about ten o'clock. It does not really get dark now at all at midnight. Everything is so new and strange to me that in these few days I have not completely learnt my way about the ship—much less learnt the ways of life; but I think it is in every way a great improvement on the *Mantua*, and I am already finding myself at home with the others although I am older than all except three.

JUNE–JULY. There is one thing which when it happens almost reconciles me to this unnatural life, and that is to see the Grand Fleet go to sea. A Spithead review in 'peace time' was a fine spectacle enough, and one to make you proud of your name and race; but the Grand Fleet under weigh is the most tremendous exhibition of power that was ever seen in the world, and I confess it brings a lump to my throat when I see it. The first ten days or so after I joined this ship we remained in harbour, and I was beginning to get accustomed to seeing lines and lines of ships at anchor as though they never did anything else. Quite suddenly came the order to proceed to sea. Away went destroyers and light cruisers in an orderly hurry, followed by more destroyers, armoured cruisers, and last of all the great battle squadrons. The way out of harbour is through a 'gate' in the submarine net through which only one ship may pass at a time, and when I say that it takes nearly two hours for the fleet to pass through the gate, some idea will be given of what a grand procession it is. Outside the harbour the squadrons form into divisions of line ahead or line abreast, and so on, and perform all manner of evolutions of which I understand only a very little. What I do understand and appreciate is the almost miraculous accuracy of line when we are going ahead —or of curve when we are altering course. Big ships like this are more than twice the length of King's Chapel (think of it!), and two lines of them can make a 16-point turn and yet keep station at a speed of 20 knots. Thinking over it in the terms of King's Chapel you will see that it takes some doing. The little destroyers are just as wonderful: they circle about the big ship at an incredible speed—dashing off this way or that if anything suggesting the possibility of submarines (generally it is a whale) is reported—and then return to heel, or rather on to the flank and ahead—like good dogs. Truly, if our people are as good with their guns as they are at handling ships, the Hoch See Flotte had better stay at home. After three or four days of these 'flaps', we come back to harbour. The anchor is hardly down before colliers come alongside and coaling begins. Everyone from the commander downwards changes into coaling rig, old football clothes or ten-year-old monkey jackets, anything—so long as it is shabby—and the next few hours is spent in being perfectly happy and as dirty as possible. It is mighty hard work for the men deep down in the collier shooting the coal into sacks, but for the most part it is not too severe. Nearly everybody can smoke—which is a thing unheard of during ordinary working hours. There are a few exceptions to the active workers, and some of these are the doctors who have to keep clean so as to attend to the cut heads and so on which occur in every coaling; also the ship's band, which takes up a position out of reach of swinging derricks and wires and blazes out ragtime and music-hall ditties. I am not sure that the band's part is not the hardest of all. There are a few ships, and soon will be more, where these things do not happen. Instead comes along a nice clean oil tank with its pipes and pumps, while the ship's company can go about its usual routine. The *Queen Elizabeth*—who is supposed to be bombarding Gallipoli but has actually been up here for several weeks—is one of these more fortunate ships, and there they have gilded a coal shovel and fixed it up on a conspicuous place, with the motto over it: 'Lest we forget'.

After coaling, the deluge. All the pumps, hoses and other things that produce water, pour a flood over the ship—inside and out—so that life for a few hours is very damp and strangely flavoured with soap. It is really only an exaggeration of the usual Saturday routine: it is a time of scrubbing, sweeping and polishing, in readiness for the morrow. And when the scrubbing is finished cabins are carpeted with old newspapers so that not a speck of dust or dirt shall offend the captain's eye when he makes his rounds on Sunday morning. That is a truly awe-inspiring function—Sunday divisions. The whole ship's company comes on deck dressed in their number one or best clothes; the officer of the division inspects his children, titivates the set of their caps or the tying of their shoes—all this while the captain goes his rounds below. Then the captain appears—attended by the senior officers, mates at arms, quartermaster, and hosts of other satellites. Everyone stands to attention as the great man progresses up and down the ranks, inspecting the backs of some and the fronts of others, criticizing the length of this man's hair or the width of that man's trousers. Truly they have brought attention to detail to a fine art in this service. All the time the ship's band plays lugubrious ditties, and I —and the other lesser fry—extract what fun we can from a somewhat boring proceeding. After divisions is church (of no great length), and then cocktails in the wardroom.

People who live on dry land are prone to consider a picnic merely rather a tiresome way of entertaining young children. Not so your sailor, of whatever rank or age. On a day of 'make and mend clothes'—or Saturday afternoon after a substantial luncheon—parties of officers go away in the ship's boats laden with supplies enough to last a week. You sail if there is a wind, or pull or get a tow from a motor-boat to one of the many coves in the island, and for a few hours you forget the cramped life of ships. Some bathe, others walk or scramble about the cliffs; but the principal point is tea. We make a blazing fire of driftwood, boil a great iron kettle (all such things in the Navy are very heavy and solid) and then fry eggs and sausages or 'bangers' as they are called. No picnic is complete without 'bangers', and it is astonishing how many of them are consumed. We eat as though we had starved for a month, and then all too soon it is time to pack up and get back to the ship.

I have had some very pleasant afternoons fishing with the Commander, and thanks to him have been able to get to some of the remoter places. We are generally at short notice for sea, and are consequently not able to go for more than an hour or so from the ship; but a commander is lord of boats and of signalmen, so we have been able to get out of sight of the fleet and still keep in touch with the ship through a watcher—posted on a hill-top. The owner of Hoy gave us leave to fish in Heldale, a deep loch among the hills where the trout are large and cunning. All this sounds rather like a perpetual picnic, but really these occasions happen only about once in ten days, and the intervals are drab enough.... My own duties in the ship are so insignificant as to be hardly worth mentioning. The ship's company number about 1240, and the average daily sick list is rather less than one per cent! We have by far the best 'sick-bay', as the hospital is called, of any ship in the fleet; and besides this there are two hospital ships up here, so that a man suffering from any sickness or injury which is not likely to be at an end within two or three days, can be immediately packed off to a hospital ship in one of the drifters which make a periodic round of the fleet. We three medical officers occupy our great intelligences on the trivial treatments of coughs and colds, cut heads and the other casual injuries of an extraordinarily healthy set of men. At a little before nine in the morning we 'proceed' (that is a very good navy word like 'forthwith') to the sick-bay, and it is a rare occasion when we have not all come away from it by a quarter to ten. A visit in the afternoon and again in the evening to see if any sick have reported—and that is our day's work. Once or twice a week we give instruction in first aid to gangs of unenthusiastic stokers, stewards, cooks, and others who would actually be our assistants in action; but these latter know what little is required of them well enough; whether they will carry it out in practice remains to be seen. We get many letters from army doctors in France telling us how short-handed they are, how they work day and night and still do not see the end of their labours, and when we hear this you can imagine how we chafe at our idle life. Of course one knows that it is all a part of an ordered plan. Whether it is well or ill ordered remains to be seen. It is presumed that a day will come—*Der Tag*—when these great ships will go into action, and the first salvo of shells that hits the mark, or the first single shell if it is a lucky one, will provide work enough for the surgeons for a day or two. But what about the second salvo or the second lucky shell? The odds are that there will be left neither ship nor surgeons. Think of Coronel, and what chance had the doctors then to deal with the hundreds of wounded before the waters closed over them! Think of the Falkland Islands, and how much work was there for all those surgeons with twelve killed, and, I think, seventeen wounded in all ships! So far as I can understand it—in talking with the professional fighters—a naval action must be a case of sudden death for one or the other combatant. There will of course be lucky ships which will sustain heavy damage and casualties and will yet escape total destruction, but they will be exceptions to the general rule. During the action itself—a business of hours at sea, not of days as on land—very little treatment of wounded can be given that cannot be adequately performed by a competent steward. When the action is finished, surviving ships that have sustained casualties can borrow from those that have not. Another point which has often been impressed on me by the professional fighters is, that a general fleet action—if and when it takes place—will almost certainly occur in home waters. That means that our ships that have sustained casualties can land their wounded at a British port within twenty-four hours, even supposing that they have not been able to hand them over to an attendant hospital ship many hours earlier. I am not grumbling unreasonably about my present life of inaction, for I know well enough that they also serve who only stand and wait; that applies to the whole of the G.F. at the present time—their day will come. But we medical folk could and ought to be better employed. The Navy is said to be the finest service in the world. The navy medical service most emphatically is not, and the chief reason of that is not very far to seek. A fleet surgeon of eighteen years' service told me the other day that he was sure he had not done eighteen weeks' real work in all his time, and judging by what I have seen of the service I cannot suppose that he was greatly exaggerating. It cannot be expected that young doctors straight from their hospitals and with any interest at all in their profession will enter a service in which there is nothing to do. When I was at hospital in London, we were advised —at about the time of taking our degrees—not to go into the Navy service on that very account, and I hope that warning will still be given until the whole medical service is reorganized.

JULY 26. We are all in a state of suppressed excitement, waiting for our turn to go and refit, which means 'leave'. The great majority of the people in this ship haven't had a day's leave since the beginning of the war, and there are even a few who have actually not been ashore in all that time.

AUGUST 9. Back from seven days' leave, with the prospect of no leave now for twelve more months. Only those who have been in prison know what it is to be let out—even with a ticket. Our ship is divided into two watches, and six days' leave was given to each watch, but as nearly all the ship's company belonged to Portsmouth it meant that most of them had the better part of twelve days' leave. My leave fell with the second lot, so I had to endure six days of a strange and ugly town; but my kind friends the Commander and P.M.O. introduced me to the Naval Club where I spent many pleasant hours, and dined in comfort. It is a good old house with a fine library and notable cellar, but it seems that there is every prospect of its early bankruptcy, which would be a sad pity as it is the only purely Naval club in existence. My six days' leave was pretty well occupied, and I even managed to squeeze in a part of a day at Cambridge for luncheon with Nixon at King's. The place is utterly changed, and my old rooms in college were occupied by a couple of nurses! It rained as it only can at Cambridge in August, and I returned melancholy to Portsmouth, almost wishing that I hadn't been there at all. Like the rest of us, I wore a very long face when we all met at dinner on the night before we sailed. It was rather like the first night at school after the holidays. But these holidays will not come again after three short months; perhaps—if we are lucky—after another year. Considering this, I think we all shook down in a wonderfully short time, told each other what splendid times we had had, and what much more splendid times we meant to have next August. We went south and north without the suspicion of an attack, and found Scapa so hidden in fog that we had to wait in the P. Firth for sixteen hours before we could get into anchorage. The newspaper folk are not far adrift when they talk about the G.F. being hidden in the northern mists, even in August.

NOVEMBER 9. (*To M.M.*) ...I have often wondered where you were, and if I had had to make a guess I should have said Campden Hill. Somehow it sounds wonderfully peaceful and remote from war, and I should dearly like to be within hail of it, although the people who live up there always make me so dreadfully shy....If, as you say, your family colony attracts Zeppelins, Campden Hill must be a more lively place than it used to be! Do you know Celia Furse? She is the daughter of my great friend Harry Newbolt, and her man's father is Harry Furse, one of the very best men living. You can't think what a pleasure there is in merely writing the names of people like that. Like Campden Hill, they are something quite apart from war, and here we know nothing of such people, for naval officers—although very jolly fellows—know only their own kind and their own job, and this becomes rather tiresome after a period of months. We talk war, war, war; damn the war; forgive me. ...I have seen the Atlantic at its most terrible and I was often horribly frightened, but I have got to the stage of losing fear of the sea, if one really can do that. I am now shifted into the largest of all battleships and am with the Grand Fleet (a name I don't like). By Jove, I wish you could see the fleet at sea steaming majestically over the sea in glorious lines! It is one of the finest things I ever saw.

At the end of last July, when I had a few days' leave, I should have greatly liked to have seen you had I known where you were. As it was I spent a few days in the country, and you can imagine how sweet the good earth smelt after all these months at sea. The trouble is that we shall not go anywhere near civilization again for the greater part of a year, unless by some miracle this bloody war (that is not swearing) comes to an end sooner; but I cannot believe that it will for it looks like going on for ever....I hate war, and all things naval and military, more than most people do, and this war has utterly spoilt all my plans for life, such as they were; but I could never look myself in the face again nor anyone else whose opinion I valued, if I did not carry on with the confounded thing so long as I was wanted. Afterwards, Heaven knows what will happen or what we shall have to live on; for myself I expect it will be the workhouse if such things still survive. What it is all for, what the use or the end of it, who knows? Forgive all this long harangue, but I cannot help it, and I can think and talk of nothing but war. It has become a horrible obsession. I do not think the people of my age will ever outlive it. I was forty last May and so I am entitled to speak like a grandfather! It is extraordinarily pleasant to hear that there is still music in the world and that you have been to hear some. How I should love to do the same thing. One begins to shrivel up mentally and morally, and it is good to be reminded that in spite of all this foolishness and horror there is still some beauty in the world. How I do go drivelling on!

NOVEMBER 10. The time slides along somehow almost unnoticed, and all the time I wish that the months would go quicker to get to the end, and at the same time I grudge every day that passes. It is such an appalling waste....For the last fortnight we have been away at Cromarty where it is easier for the ships' companies to get ashore and play games. I have had two or three good walks, and the autumn colour of beeches and birches has been particularly fine. The Rear-Admiral in command on shore is Pears, and it would have been pleasant to go to his house, but of course a surgeon is very small beer and does not go and leave cards on an admiral. I have just had sent me a copy of the *Cambridge Chronicle* containing a list of more than 550 Cambridge men killed or died. King's for its size seems to have lost more than any other college, and some of the most promising of the younger men. When all these good people are going it is hard to sit still in the security of a battleship, with a daily lessening chance of doing anything active.

DECEMBER 13. I would sooner sniff the reek of wood smoke or dead leaves than that of any of the seven seas. It is now exceedingly cold and much more snow than usual; the hills are all white and look almost mountains. Nobody gets ashore nowadays and it is too rough for small boats. My bulbs are coming up all right and the Roman hyacinths will be flowering in a few days. Several other people on the ship have bulbs and we ought to have a good show early in the year.

I was very annoyed at seeing my photograph in *Country Life* the other day. H.N. had no business to do it and I wrote and told him so....

DECEMBER 15. I was informed this morning at 6.15 a.m. that I had been appointed to the *Vengeance*, and must report myself to the Admiralty 'forthwith'. A drifter was alongside at 9 a.m., so I had to bustle, but was just ready in time. I was awfully sorry to leave the ship as I like the people very much, and was quite prepared if not content to stay in her to the end of the war. Could not make out why they had shifted me.

On the 17th I went to the Admiralty and was informed that the *Vengeance* is going out to East Africa in connection with operations against German E.A., and that H. Newbolt and Ralph Vaughan Williams had separately told Smith-Dorrien that I knew all about E.A. and its diseases (which I don't), and that he, Smith-Dorrien, ought to take me. So Smith-Dorrien asked the Admiralty to send me, and this they are unexpectedly doing. I am told that they are going to land a number of marines and guns to go up with the military force, and that I am to accompany them as M.O. It is a splendid stroke of luck for me, as I am likely to be much more useful there than with the Grand Fleet, and I may even see some fighting. The *Vengeance* is full up and cannot accommodate me, so I am being given a passage in the Union Castle S.S. *Saxon*.

DECEMBER 19. (*To Henry Newbolt.*) You have indeed heaped coals of fire on my head! and I must crave your forgiveness for upbraiding you the other day. I was dug out of the *Agincourt* at a moment's notice and told to report myself immediately at the Admiralty, not knowing in the least what it was about. They appointed me additional to *Vengeance*, the flagship of the African squadron, and I am to have medical charge of the naval landing party. They are going to land guns from the ships, so we ought to do something useful. It is just the sort of thing that will suit me, and I am immensely grateful to you for suggesting my name to General Smith-Dorrien....Blessings on you....

JANUARY 14, 1916. (*Cape Town. To his Mother.*) I was to have gone on to Durban but have become temporarily attached to the Army. General Smith-Dorrien has been seriously ill all the voyage out and is only now beginning to improve. I helped the ship's doctor to look after him on the way out and we got on all right together, so he (S.-D.) asked me to stop and look after him, and here I am with his two A.D.C.'s and a nurse in this hotel. It is rather an anxious job for me, but of course it is a compliment to be employed in this way. The *Vengeance* is likely to be a good deal delayed and will not get to Mombasa before us, so I am really not losing any time; in fact but for this I might have kicked my heels about Mombasa for a month....Such a blue sky here, I wish you could see it....

JANUARY 27. (*Rust en Vrede, Muizenburg. To his Mother.*) We came here this afternoon by motor from Cape Town, about twelve miles. Sir H. has not got on nearly so fast or so far as we hoped, and it is doubtful if he will ever be fit to get to East Africa. That is a horrid question which will have to be settled during the next few days.

This is the swagger watering-place of South Africa on the east side of the Cape peninsula, and this house is the 'bungalow' of a millionaire fellow called Sir Abe Bailey. It was built for Cecil Rhodes by Baker, the man who is planning

New Delhi with Lutyens. Dutch style—like many of the good houses in this country—and full of really good old Dutch furniture with appropriate prints and china; a cupboard of pewter would make your mouth water. It is an extraordinary contrast to my occupation of the last eighteen months, and for the first time since the war began I am really earning my pay by work: before, it was the bread of idleness, but very hardly earned. Staying in this house is Dr Jameson of the Raid—a most interesting and attractive person.

JANUARY 31. The General is very bad and we have had to send for a surgeon from Cape Town. I gave chloroform this evening. He improved rapidly afterwards, but it is not possible for him to continue his command in B.E.A. and General Smuts is appointed in his stead. Great disappointment to everybody.

General Smith-Dorrien, upon his return to England, wrote to A.F.R.'s father saying: 'I want Mrs Wollaston and you to know that I left your son well, but chiefly to let you hear from me how deeply indebted I feel to him for the devoted and skilful care he took of me through a very serious illness. The simplicity and honesty of his character made him a very pleasant companion, and it was with a very sincere regret that I left him behind. All my staff too were devoted to him, and for myself I am looking forward to the time when we shall meet again...'.

FEBRUARY 21. Landed at Kilindini and went on board the *Vengeance* to get stores, etc., from P.M.O. Left by train with the remains of a S.A. contingent, and was disgusted at a wayside station to see the S.A. soldiers jump out, climb all the coconut trees they could, and loot some several hundred nuts without any protest from their officers. These men have no discipline at all.

FEBRUARY 24. Voi in the early morning, then Maktau. Stayed at Maktau for a fortnight and then went forward to Mbuyuni railhead. There we found the 10th Marine Battery.

This is a large camp not far from the Front in this region. Our detachment is two batteries of marines. It is not at all a bad place—very hot by day and quite chilly at night. Water comes by pipes from hills some miles away. There is none here and so no mosquitoes, which is a very great blessing. We are a bit raised up from the surrounding plain—which looks in some lights very like the sea, yet actually it is covered with scrubby trees. About 20 miles away are low mountains, with beyond that the glorious mountain which the Germans have no right to. We shall take it from them. One day I am going to ask General S.-D. to give me facilities for climbing it.

MARCH 7–9. Forward 5 miles to Serengeti Camp.... Later shelled Saleita Hill. This was the hill where the S.A. Infantry and others took a nasty knock three weeks ago. Now we have a large force against it and about twenty guns which make a fine noise and a grand sight to watch. There may have been a small force of the enemy on the 8th, but there was nobody left on the hill on the afternoon of the 9th when the S.A. Infantry advanced to the top.

MARCH 10. Moved forwards past Saleita Hill and went 14 miles to the Lumi River. Camped in a very horrible swampy place in the midst of trees, where we were in much danger of attack. Next day river bridged and we camped at Taveta. Watched the battle of Latema and Reata until late at night. Guns shelled German position on Latema Hill 3 miles to west; infantry attack after dark. The shells set fire to the bush on the hills and many poor wretches were burnt; our casualties about 200, and the enemy's much greater. The 9th battery (Russell's) was in action, but the 10th did not fire.

Stayed at Taveta, and on March 18 went forward to Himo River where was advanced G.H.Q. and a large force. During the next week a lot of fighting in the flat country below, as far as Kahe, whence the enemy finally escaped. After this,

G.H.Q. went forward 19 miles to Moschi; we remained at Himo. I look after Rhodesians, as their M.O. is sick in hospital.

APRIL 30. (*To J. H. Clapham.*) ...Nixon's death was in no way a surprise to me, but he is a very great loss; you know I was very fond of him. ..It is only to-day that I have discovered that I have no friend (yes, one only) who is not older than myself, and I have not made a new one who counts at all for many years—so I must cherish those that are left.... We are now camped for the rainy season on the slopes of Kilimanjaro where we shall probably spend another month. It is not exciting. It is a vilely unwholesome region and there is much sickness, so I am pretty well occupied. Disease is much more deadly than the enemy out here and I am glad to be of some use. K. is a wonderful fine mass and I hope to have a chance of getting to the top before we clear out. When that will be goodness knows, but I don't think the Germans can last many months. They might hang on in small parties for a long time, as it is a huge country and very difficult to traverse, but they have been cut off from the outside world and many necessaries of life for some long months now, so I doubt if they have much heart left for anything. But give the Devil his due, they are very fine fighters in these parts, as elsewhere. We get very scanty news from outside: occasional unintelligible Reuter, and papers two months old. However, Africa is a fine school for patience, as I learnt long ago; but Africa in war time with an army is very different from Africa in peace time, and I care not how soon I leave it.

Am very sad about old Nixon's death. He was one of the kindest friends I ever had, and it means almost the end of my connection with King's, where I have been made welcome for twenty-three years—a big slice out of my forty-one. It is indeed a quickly changing world.

I am in medical charge of a large batch of artillery. Malaria and dysentery are the commonest disorders—very often contracted by treating Africa as if it was Europe. Another cause of trouble lies in the rain which comes down solidly, making the roads impassable and the men necessarily inactive; they consequently go sick. In the matter of companions the *Agincourt* was greatly preferable to this. There were several people there whom I liked greatly, but here there are none whom I take to much.

MAY 2. There is a chorus of frogs singing outside my tent and it is after ten o'clock, so I must think of bed as 'sick parade' begins at 6.30 a.m. This is peony time in England; how I wish I were there to see it. Wherever I am, I always want to be in England in May and June.

JUNE 12. Left Himo. Went to Taveta and then to Mbuyuni. Joined the 15th Heavy Marine Battery. This battery was to have started forward this week but now the order has been cancelled....Most disappointing.

Very little doing and pretty deadly it is. Everybody except a few artillery batteries has gone forward to chase the enemy. They are making it as far as possible a South African show, and the Imperial troops—ourselves included—are kept in reserve. It is very galling, especially as we know how different things would have been if General S.-D. had been in command. Here I have about 300 men to look after, but on the whole they keep very fit and there are less than 10 per cent. sick at any time. Food very monotonous; no vegetables but onions which I have learnt to eat but without any liking. For a few weeks after the rains we got some very good green leaves known to the natives—almost as good as spinach.

AUGUST 8. (*Mbuyuni.*) Have joined No. 15 Battery and we have been on the move for some time towards the middle of ex-G.E.A.—about 250 miles from where we started. Pretty hot, and dust beyond belief. Oxen and motor lorries on soft roads make a fearful mess but we have got through without great difficulty. There is a very big movement going on, and the Germans have probably gone back beyond their railway line.

AUGUST 10–14. Camped at Kangata, and on to Lukigura—a most foul and stinking camp. Dead oxen and mules and horses all along the road. Arrived at Msiha. The whole place honeycombed with dugouts....Went 12 miles south to Russongo River, and then on to Twnani where are advanced G.H.Q. and most of the artillery. Here we really are at the 'front' except for infantry.

AUGUST 23. To Dakawa and then east along Wani River. Next day went south-east and down left bank of Ngeri River. Slight fever; first I have had in this expedition. On September 4 we hoisted British flag at Morogoro. A few days later I was appointed to Royal Flying Corps, 26th Squadron. Still keep on 10th M.B. and 134th M.B....

NOVEMBER 12. I went down to Dar-es-Salaam a month ago for a week on duty, and got over to Zanzibar for a day. I thought Z. a most interesting and beautiful place. I walked in gardens of cloves and ate fresh fish and oranges. That last doesn't sound very exciting, but it is after months of trek ox and no vegetables. The change did me a world of good for I was getting rather 'mouldy', but now I have taken a new lease of life mentally and can regard the prospect of many more months of this more or less equably. But I often wish I were back in the Fleet.

NOVEMBER 24. (*Morogoro. To Henry Newbolt.*) ...It is an age since I wrote to you and more than an age since you wrote to me, but I will forgive you, as you are no doubt better occupied than I am, and will wish you and yours all happiness in the New Year; may it see peace! That is really all I have to say—just a handshake and a dirty one at that. There is a devil of a wind blowing and filling my hut with a fine red dust. We pray for rain which is due, and at the same time we hope that it may not come and put an end to any operations that may be in progress....On the whole it has been a very unsatisfactory business and the British units of us have sighed a thousand times for Sir Horace. We have covered a wonderful extent of the most damnable country in pursuit of the retreating Hun, who has now escaped into swamps and other frightfulnesses to the south of here. We are about ten to one, but still he slips away. Of course the end is certain, but the cost is great. Our people go sick in thousands and die by scores, so the likes of me are kept busy and consequently fit. A month ago I went down to the coast on duty for a few days; it was good to smell the sea again, and better to find a ship or two where I was hospitably entertained and forgot for a moment the little jealousies of the land forces. I have learned something of both these Services since 4.8.1914, and bow my head to both of them, but I shall always be glad that a chance word at dinner at the Savile on August 3 made me go to the Admiralty rather than to the W.O....

DECEMBER 20. (*Morogoro.*) We are now off again in pursuit of the errant Hun. I hope we may really mop him up this time, but we have had so many disappointments before, that one cannot feel very sanguine. The next rainy season is not very far off, so if we do not finish it off now we shall stop several months longer, which heaven forbid.

JANUARY 9, 1917. Left Duthumi. Went through Dakawa, Kissaki and Wiransi, where I caught up the 15th Battery and slept at Beho-beho. Next day reached Rufiji....We shifted camp on to ridge north of camp. Some firing on both sides of river....The Germans are on the south (right) bank....

JANUARY 18. Advance beyond river begins. I join hospital line for the time. Awful scrimmage for the next few days; much work with wounded and sick.

FEBRUARY 26. Left Kibamawe. Very hot walk to Beho-beho and to Tchogowali. Saw many hippo in salt lake and plenty of game. Next day went to Wiransi and on to Morogoro and Dar-es-Salaam. I find my new job is to run the sick in two ships—the *Pemba* and the *Mafia*—from Rufiji to Dar-es-Salaam. Not much of a job, but I will see.

MARCH 30. (*Rufiji River.*) Have been put in charge of a small ship which goes up the Rufiji river from the sea to a post about 100 miles inland, and comes back laden with sick. I went up for the first time three weeks ago and brought back about 100 sick; now we are on our way up again, but at the moment we are stuck on a sandbank and are likely to remain so for a day or two. It is a very small ship (H.M.S. *Hyacinth*) and 250 people on board—so the noise and smell can be imagined. We are packed like herrings in a barrel! There is not much to do until we turn round and come back, and then I have my hands full enough. After this particular job I have no idea what they will do with me. My fear is that I shall be sent to a ship on this coast, but I would sooner be with a job on shore here for twelve more months than do that. This is a great river in full flood and the country on either side is under water for miles. There are scattered huts along the banks, or rather where the banks should be, and they are raised high on piles—the people moving about in dugout canoes. Except for the heat and the strong current it is not unlike the Norfolk Broads.

APRIL 1. We are stuck on a wretched sandbank discharging cargo into boats and lighters, and are not likely to get off before the day after to-morrow. What wouldn't I give to be back in England now! There must be primroses and violets and chiffchaffs at home.

MAY 16. Discharged our cargo of Germans at Tanga; then back to Dar-es-Salaam. Was feeling glad to think that this was the end of my *Mafia* job, when I found a note telling me that the *Mafia* was going to Lindi and back to evacuate sick, and that I must go. Went on board without my boy and in a great hurry; very much annoyed. Went ashore at Lindi. Pretty harbour and it is good and cool here at night.... Walked uphill towards the watch tower, but felt the sun too much so came back and lay down. Next day entered Mohoro river, and then went in boat to where the sick from Mohoro were waiting. Collected 152 sick and sent them down in boats, and so back to Dar-es-Salaam.

JUNE 2. Am appointed to Lindi as Port M.O. and in charge of sanitation. Got my little dog 'Sally'. She was a passenger from Kilindini—unclaimed—such a little dear, full of life and very fond of master. I can't think what I have done all these years without a dog.

JULY 27. It looks quite possible that Russia may make a separate peace, in which case I don't see that we can ever bring the war to a really successful end—even with the huge American armies of two years hence. Wouldn't it be better to stop it now before there is any more death and destruction? I can't help thinking that the mass of Huns are as tired of it as we are. In any case I hope we are not going to fall out with Holland, for with all their faults they have had a very difficult part to play and must have suffered greatly.

OCTOBER 21. (*Lindi.*) The Admiral has given me permission to go from here when I consider my work is finished. That should be in January, for by that time I shall know whether or not my drainage scheme is working properly. I have been trying to make the place as free as possible from mosquitoes which are very bad here in the rains. I have been nearly two years away now, and shall be glad enough of a change of climate, for one loses energy and one's mind becomes a swamp out here. Nevertheless I have been well occupied and have kept fit—which latter very few people can say.

JANUARY 8, 1918. (*Lindi.*) The Germans have been driven out of German E.A. and are touring about Portuguese E.A. There are not very many of them left, but they are doing a lot of damage and will doubtless carry on for months. Lindi is still the most important base of operations, so I am staying on here. We know almost nothing of the war, except that there is chaos in Russia and that there are talks of peace; how I wish it would come.

FEBRUARY 11. I was astonished to hear yesterday that I had been awarded the D.S.C.* For what on earth I cannot imagine, unless it is a sort of consolation prize for being the oldest R.N. surgeon.

FEBRUARY 26. (*To his Mother.*) ...I am so sorry you have the impression that I don't like people; that is quite wrong. I have found so many people out here that I like, and many I am sure who like me, but one does not make many new friends at my time of life, especially in these circumstances, where people come and go so quickly, and die or are killed. I have learned to like many people whom aforetime I should have taken no notice of at all; indeed, I find that most people have their good points and I honour them accordingly—especially those who can show their better side in these conditions. No one knows—and no one probably will ever know—what the rank and file of Europeans have had to put up with in this campaign; but that is a censorable remark and that is

* The official announcement was: 'For conspicuous devotion to duty and for his unfailing care of the sick and wounded during military operations in East Africa. During the operations in the Rufiji River Delta, he voluntarily attended to the casualties of the Rufiji River transport service in addition to those of his own unit'.

one of the reasons why my letters from here have been so bald.

MARCH 21. (*Naivasha.*) Left Lindi on March 2. Such a relief to get away, and as I thought before, this place is the most beautiful in the world. Indeed even more beautiful now, by contrast with all the hideous places I have seen these two years of war, and also by its perfect peacefulness. Here there is no sound of war and I could wish to stop for always. The birds and beasts swarm as ever and the mountains and lake are beyond words beautiful. I can shed more than twelve years of my life and find myself back in 1905, sitting on the sunny slope of the island here, where hundreds of willow-wrens sing in December as perfect a song as in an English woodland in June. I could have bought this island then for a thousand rupees, if I had had so much; now the present owner refuses eight thousand pounds and is likely to get ten thousand. What an opportunity lost!

At the end of June 1918 A.F.R. was back in England. He was given a month's leave, and this he spent at Capel Curig.

Fished in lake; no fish but very pleasant. I walk about and get wet and am happy so long as I can forget what I go back to to-morrow fortnight. How I hate the prospect of it all.

AUGUST. Joined H.M.S. *General Wolfe.* She is a large ship of the small sort with a very few officers. The captain quite nice and inclined to be friendly. Am afraid I shall have to rely a good deal on myself, which is a pity, as I do not think my own society is the best or even sufficient, though I have had to put up with a good deal of it in past years. I can't say that I rejoice at being cooped up in a ship again but I have been mighty lucky to have kept out of it for so long.

I have got a very pleasant book called *Small Talk at Wreyland,* written by an old Devon squire—Bovey Tracey way.

All about anything, people and customs, cider and his garden, the neighbours—and nothing at all about war; a very placid sort of book.

In September 1918 A.F.R. was appointed to the *President* for special service in the Naval Intelligence Department. Six months later he joined H.M.S. *Fox*, and sailed for Arkhangel in April 1919.

MAY 17, 1919. (*Arkhangel.*) The last six days have been spent struggling through the ice at entrance to White Sea. For three days we made no progress at all, but finally got out with the help of ice-breakers. A very interesting experience and new to me. It was a glimpse of the real Arctic which I would not have missed for much, but enough to show me that I had done right in avoiding Polar expeditions.

MAY 27. Bitterly cold and snowing. Not a sign of spring so far. I suppose it will come with a rush. There is no local news and I do not know what our movements are to be, but I rather hope to get shifted out of this into some other craft that goes up river.

Troops landed. Sleep on board *Borodino*....Willow-wrens, redstarts and white wagtails, a swallow, oyster catchers and tufted ducks.

JUNE 10. (*Dvina River. To his Father.*) The White Sea is not yet clear of ice, and ships are occasionally jammed for days at a time in spite of ice-breakers. Arkhangel looks very picturesque from the river, many churches with golden onions atop—but it is a dirty place when you get there. The streets are swamps, with stinking drains on either side covered with badly fitting planks on which you walk. Nearly all the houses are wood-built and nothing to look at. The town ends abruptly in rubbish heaps, beyond which is 'tundra' (mossheath, dwarf willow and swamp) where it is almost impossible to walk.

Before we left Arkhangel about 5000 men of the new force

arrived. A splendid set of men, numbers of V.C.s, D.S.O.s, M.C.s, etc., in the ranks. They made a formal landing, and were received by the local government who presented the general with a large cake and a bowl of salt. Now we are about 200 miles up the Dvina in a small paddle steamer called *Borodino*, which is the temporary headquarters of the naval commanding officer of the river expedition. I am temporarily attached to his staff, but do not expect to stay long as there is not enough to keep me occupied. At present we are within long range of the Bolos who hurl ineffectual shells at us, but we shall probably drive them back soon and begin the voyage which I hope will bring us to Petrograd—about 1300 miles from Arkhangel. The country is quite flat—woods of fir and birch and swamps—much the same as Finland....Birds very many, but I have not had many chances of getting ashore, and the mosquitoes are already a great plague. No fishing, and no prospects of any if I stay on this river....

JUNE 15. (*To Henry Newbolt.*) I was 44 a week or two ago, but I found this morning that I am not yet too old to blush for pleasure that an old friend has so high an opinion of me. But you pay me far too great a compliment. You would put a pygmy among the great ones of the road, and he is not fit to be seen in such a glorious company.*...I am amusing myself at odd moments in writing letters of a naval surgeon '14–'19, but I doubt if they will ever get beyond the pencil copy. It is sad to miss another English summer, but there are compensations here—rivers and forests and occasional gorgeous days.

JUNE 21. Joined H.M.S. *Humber*. Moved up river beyond Tulgas church. We have daily duels at long range with the Bolo gunboats who make quite good practice but miss us

* A.F.R. refers to the inclusion of his explorations in Sir Henry Newbolt's *Book of the Long Trail.*

narrowly; possibly we do the same. Our advance up river is hindered by mines which they have laid in the channel. They make a splendid geyser when they are blown up, and kill a lot of fish which are eagerly grabbed by ships as they float downstream. This ship is one of the river monitors built for Brazil in '14; about 150 men and 9 officers....If the river continues to fall at its present rate we shall not be able to get much further, and it is quite possible that we shall not be able to get down until the water rises again at the end of August—or thereabouts.

Day after day brilliant sun—quite hot—and we are wearing white uniforms. The pity is that we can never get ashore. Mails are very bad, and news of the outside more scanty than it was in East Africa. Yesterday we heard that the Huns had sunk their ships at Scapa. They certainly have the laugh of us there. I suppose we are serving some useful purpose here but it is not easy to see without a full knowledge of the whole Russian situation. I wonder who has that?

JULY 8. Up river to Troitsa....Big bombardment with good result; Bolos retreated.

There is not now the smallest possibility of our getting through to Petrograd, and for the moment we can go neither forward nor backward by reason of the lowness of the river and shortage of supplies. The Russian troops have been very disappointing and are not to be depended upon at all. If it were not that we should condemn Arkhangel and the people who have helped us to an awful fate, the best thing we could do would be to clear out and leave Russia to mind her own business.

AUGUST 10. (*Troitsa.*) Bombardment by flotilla. Opposition by Bolos. *Humber* fired 250 rounds. Tremendous attack by land, water and air, and we got everything we wanted except the Bolo gunboats which escaped up river. We took over 2000 prisoners and practically all their land guns, so they are not a very active force now.

The river has risen about 3 inches, but it will have to rise another 3 feet if we are to get the ships away. It would be very ignominious to have to sink them here and go down river in barges.

SEPTEMBER 11. Left Arkhangel and got into open sea.... Home to England—and demobilization?

OCTOBER 23. Demobilized to-day. Laus Deo.

VIII

FIRST MOUNT EVEREST EXPEDITION

To his great delight, in November 1920 A.F.R. was elected to a Fellowship at King's. This appointment to a Fellowship was a unique tribute to his character, for although learned in the far places of the earth and in his knowledge of men and things, A.F.R. was entering King's as a plain man among scholars. 'Notwithstanding the honour I feel has been done me', he wrote, 'I cannot but be amused when I think of Henry VI turning in his grave at the thought of a man with a poll degree getting a Fellowship'.

At King's A.F.R. soon found much to occupy himself with. He was not only keenly interested in everything to do with the College and its buildings, but he became a great lover of the plants and trees in the Fellows' Garden and College grounds.

MARCH 5, 1921. (*To his Father.*) Did I tell you that I am a good deal interested in the trees in the backs? This College owns nearly all the trees there—as far as St John's Wilderness, and we are responsible for their safety and welfare. In the last forty years or so an immense number of them have been blown down or have been badly mutilated by gales, and now many of those that remain are unsafe and ugly, so we have decided to cut down a number of these relics between our back gate and Clare gate, and replant. There is a good deal of outcry as you might expect, but I am convinced that it is the right thing to do. Attempts had been made to plant in the gaps made by fallen trees, but the young ones have no chance of growth. The place will look rather bare for a hundred years or so, but what is that in Cambridge?

Much can be done to improve our own Fellows' Garden, for we have a number of lovely trees and shrubs needing kindly attention, but they stand at present jostled up against a lot of uninteresting shrubs that have no business in anybody's garden. It is astonishing how much the beauty of a place is enjoyed and how little is done for its future benefit.

Hardly had he time to settle down to the life of a College don when, in February 1921, he was invited to join the Reconnaissance expedition to Mount Everest as medical officer and naturalist. This expedition was led by Lieut.-Colonel C. K. Howard-Bury, and the other members of the party were: Harold Raeburn, Dr A. M. Kellas, G. L. Mallory, C. H. Bullock, Major H. T. Morshead, Major O. E. Wheeler, and Dr A. M. Heron.

MAY 14. (*Darjeeling. To M.M.*) ...We are busy packing things to make mule loads for our start on the 18th, so this is the last opportunity of writing to you before we get into Tibet. There is a glorious view of Himalayas from here, and I have seen glimpses of Kanchenjunga towering over the clouds, more than 28,000 feet up. I am sure Everest will not be half so fine. I wonder what riding on yaks will be like, and are they all they are said to be in *The Bad Child's Book of Beasts*?...In spite of being 'a strange man of some twelve years' acquaintance' I am going to write to you if you do not think me a bore. I think you are dreadfully unapproachable, for you have so much family and you all know each other and your friends so intimately that a stranger like myself has no point of contact at all. And then you live in London where everything is so unnatural. I can, at least I think I can, learn much more about a person under the open sky than in all the drawing-rooms in London. But then I am really a savage at heart and I hate crowds and cities....You say that some day you will use a pick-axe to me. Well, I take it rather as a compliment, implying that you would like to know more than my physiognomy. I wish you would try it, but of course I cannot guarantee that you will find anything worth finding! When you begin to pick you must make allowances for the quite abnormal life I have led for a long time in uncivilized places, where I have always had to be content with the company of men....I wonder what I really seem to be to you? Probably rather a forlorn old buffer, (I shall be 46 on the 22nd of this month), with an interesting past and a very uncertain future! That is really a pretty accurate description. The future is what really rather frightens me...but enough of this stuff—and may all good be with you....

MAY 18. Wheeler, Mallory and I left Darjeeling to-day with mules and coolies. We rode on ponies, not bad little beasts, and went uphill to Ghum, then slowly down for about 15 miles. Cloudy and no views; forest and occasional plantations.

It was a pleasant ride from Kalimpong to Pedong, 4700 feet. Fine butterflies on the road and wonderful *Datura* plants in flower. We met many parties of little mules laden with wool coming from Tibet. They have strings of bells on their collars and make a pretty music as they go along.

Down steeply to Rishi river—boundary of Sikkim—in sweltering heat. Much cultivation of rice in terraces, very like Java. Over a ridge, and then down to the Rongli river. Spent the night here.

MAY 22. My 46th birthday. Cannot move on to-day as the mules need rest. One died yesterday, several have been left behind, and many have sore backs. They are very fat and soft, and the wet weather and heavy loads have galled them badly. Bears very common here and they do much damage in the maize and rice fields. The natives make little platforms or 'machans' at the corner of the fields—8 to 10 feet high—with a grass roof and fireplace where they sit up at night to protect the crops. The cicadas here of most penetrating loudness.

Our bungalow is distinguished by having a 'library' of old magazines and tattered novels. This kind of travel with a dak bungalow at the end of every day's journey is a great boon. No need to clear ground for a camp, and no pitching tents in heavy rain. You just walk in and find furniture, knives, plates, etc., even lamps and oil! A man called a 'Chowkedar' is in charge, and you pay one or two rupees a day.... Cardamoms cultivated here.

MAY 23. A fine morning at last, and so we were away before 8 a.m. Went steeply uphill through the Rongli valley, keeping fairly close to the river which is a fine torrent most of the way. Beautiful forest, fine walnut trees, buddleias, and many butterflies. We go by steep zigzags to avoid the cliffs, and the road is made of stones set on edge—a sort of causeway—for the ground is mostly a stiff and greasy clay which would make going impossible. Met hundreds of Tibetan mules coming down with wool. Got up nearly 5000 feet to Sedongchen (7000 feet), and here I saw for the first time a Lammergeier. As the worst part of our road lies ahead now it has been decided to send all our mules back and substitute locally hired mules, of which there seem to be any number over this pass. Our own animals appeared at first to be very fine, but they suffer badly from being out of condition.*

MAY 24. Up to-day along very steep hillside above Sedongchen, through fine forest full of flowering shrubs, ferns and orchids. At about 8000 feet began rhododendrons, at first arboreum and argenteum (mostly over), and then a lovely little apricot-coloured one which I think is Cinnabarinum; it seems to be occasionally parasitic. Begonias at about 8500 feet. At about 10,000 feet begins the great show of rhododendrons, truly a most wonderful sight, with flowers of every shade and shape and size. I must get seeds of specimens on my way back if I can find them. Saxifrages, yellow poppies, irises (not yet flowering), a blue poppy, a little yellow rose,

* These commissariat mules were very good on the plains, but quite unsuitable for mountain work.

a white ranunculus and a tiny little primula—all at 12,000 feet.

Went along the hillside more or less horizontally, and on to a shoulder of the mountain—bare of trees and covered with a deep blue (almost black) primula with a yellow centre.... Up and down through open ground, with here and there remains of snow avalanches. At 12,000 feet was the village Gnatong—a wretched collection of wooden or mud huts with shingled roofs. Swarms of mules come down from Tibet, and pigs and fowls mess about in the mud—all rather like a very dirty Irish village.

Made a fine fire in our bungalow, and temperature was at 50° at 2 p.m., which feels cold enough to me. House martins are nesting under the eaves of the bungalow.

MAY 25. On uphill to-day, rhododendrons diminishing in size as we mount up, and at 12,500 feet a *Fritillaria*. On to Jelep La (14,300 feet), the boundary of Tibet. Loose stones, mud and grass. We saw the remains of a wall built by Tibetans across the valley to keep out invaders. Went very steep down from pass, in clouds at first, then passed a small tarn where trees began. Pines and rhododendrons again. As we went lower the sun came out in a quite different sky from the Indian—of a much more home-like and familiar blue. We followed the course of a real mountain torrent, crossing by a plank bridge from one side to the other to avoid cliffs. The vegetation now more European, with oaks, walnuts, mountain ash, many beautiful primulas, a bushy jasmine, cotoneaster, anemones, and all manner of other familiar looking things whose names I don't know.

Came to Richengong, a well-built largish village much like a Tirol village. Here we joined the Chumbi valley and came down to the banks of the Chu—a wide swift river about 50 yards across, fed by snow. The country is gay with pink spiraeas, yellow berberis, and a fragrant white-flowered bog-myrtle. Found a very beautiful *Enkianthus* with clusters of pink and white flowers. Cultivated fields—potatoes and buckwheat. At Yatung we had two very pretty *Cypripediums* and a red bee-orchis on the tea-table of our bungalow.

MAY 26. (*Yatung.*) I wandered about the village this morning and thought it a squalid little place, but the people look contented enough, though very dirty. Walked up the lovely Kambu valley and found a little aster very like the Alpine aster. This whole valley is scented with roses, and the red one is a most sweet briar. Many pretty cut-leaved maples and sallows. The river forks about 2 miles up and I followed the northern branch which was milky white. At the head of it are hot springs famous in these parts for curing all the 440 diseases known to man. The other river comes from Tang La and is quite clear. Saw an Alpine chough, a *Phylloscopus*—with a song very like a wood-wren, and another with a song like a Mandelli's willow-warbler; a hoopoe and grey-backed shrike.

MAY 27. Up the open Chumbi valley to Galinka, and through a fine rocky gorge to Donka village, very well situated over a shoulder of mountain, barring the lower valley from a level plain—evidently an old lake basin above. From here we went across the Lingmatang plain where a number of yaks and goats grazed on the young grass, and in a month or so they say it will be carpeted with flowers. Our ponies had a nice scamper over it.... Up through birch woods and juniper trees into a lovely valley, better than any I have seen in the Alps. Our path crossed and recrossed the rushing stream, winding under cliffs and steep slopes, among sweet-scented cherry, pretty prunus, sweet yellow primulas, small narrow-leaved pale irises, many rock roses, and strawberries in masses. Before getting to Gautsa I saw a dipper in a stream at 12,000 feet; white-capped redstarts were common. It is impossible to exaggerate the beauty of the scenery and vegetation.

MAY 28. Still up the valley. Trees now rapidly dwindling, and after a few miles none were left except very dwarf rhododendrons which covered the hillsides like heather. I heard cuckoos calling at 14,000 feet....The stream—here called Ammo-Chu—grows smaller, and the hills become rounder—very like the rounder hills of Sutherland and Caithness. Amazing change in such a short distance. I saw dippers with young (grey) in a stream.

A yellow primula covers the ground more thickly than cowslips do in England, and the air is laden with its scent. Many blue poppies, and a white anemone with five or six flowers on one stem.

Round a bend we got a splendid sudden view of Chomolhari, filling the end of the valley and not unlike the Matterhorn. Then out on to open rolling plain where the upper waters of the Ammo-Chu meander like a bog stream through what must be at some season a bog. The yaks, grazing all about the plain, like to scrape holes in the ground and roll in them; the calves are frisky little beasts. Our ponies galloped now and then, for they were glad to be on soft ground again. The earth was riddled with burrows of a jolly little fat tailless mouse, and they scampered into their holes in swarms as we came along....At Phari-Dzong crowds of people came out to see us. It was a small village of mud houses, with a fort on a hill occupied by the Jongpen who came to visit us in the afternoon with a gift of half a sheep.

MAY 29. (*Phari-Dzong.*) The second party arrived this afternoon. Dr Kellas not at all well. I am afraid some of us already have our insides all wrong. Certainly I spent half today in bed having to starve myself. Our cooks are thoroughly bad and our food has been very bad all the way.

MAY 31. Left Phari-Dzong with a wonderful assortment of transport animals: mules, very small oxen, and miniature donkeys which walk along at a splendid pace, with bells dangling at their throats and carrying the most astonishing loads. It was a gentle rise of about 800 feet to the Tang La, then down an almost imperceptible slope over open desert country to Tuna. Infrequent plants, but I saw a purple *Incarvillea Younghusbandii*, with leaves below the ground. Heavy snowstorms on the hills around us, but we escaped with only the bitter cold wind (luckily south) on our backs. Herds of Kiang (Tibetan wild asses) are to be seen on these plains, and a few gazelle; birds almost absent. We rode on mules, and Kellas was carried in a chair. It was a ten hours' journey but he was not much the worse for it.

JUNE 1. The wind did not get up till about twelve o'clock, so this morning's trek was very pleasant. Beautiful views of snow peaks beyond the plain on our right. We stopped by a wonderful spring—bubbling out below a limestone cliff—which became a decent sized small river in a few yards. Here were Brahminy ducks and black-headed gulls, a yellow-headed wagtail, and a small frog. Seven miles to Dochen, on the side of the big lake called Bamtso. It was shallow, with patches of weeds, and there were wonderful colours as the wind rose. The sky has been blue all day with white clouds, and over Bhutan and southwards we can see the heavy white monsoon clouds which lie over the lower country.

Here is the last dak bungalow we shall be in, for we branch off the Gyantse road to-morrow.

I walked down to the lake and found many bar-headed geese, ringed plover, and terns. Redshanks nest here, and families of ruddy shelducks and garganey teal swim on the water. I watched a red fox stalking a pair of geese, and had the satisfaction of saving the birds and spoiling his game. Both here and at Phari and Tuna quite a lot of barley is cultivated; presumably it ripens. Potatoes are grown in the Chumbi valley up to 12,000 feet. People make a very intoxicating liqueur at Phari and other places and lots of our folk have got drunk on it.

JUNE 2. To-day an 11 mile march over Dong La, 16,400 feet, and then down a long stony slope to Khe, where we pitched our tents for the first time. Water very bad.

JUNE 3. Kellas very unwell, but no dysentery; pulse very poor, and he tells me that he has recently lost a stone weight, which accounts for a good deal of his present weakness. It is impossible to stop anywhere until we reach Khamba Dzong.

Camped in a curious fold in the hills where there were about twenty Tibetan families camping with hundreds of sheep, goats and yaks. They live in black tents made of very thick yak's hair.

JUNE 4. Along the same valley as yesterday, over a spur, then up over a long pass (17,100 feet). Kellas goes very slowly, riding on a yak. At about 5 miles beyond pass I was told that Kellas had broken down just short of the pass, so I went back and we carried him to a Tibetan hut where a woman made hot water and we mixed bovril and brandy from my saddle bags. Then, with six men we carried him very slowly about 10 miles—stopping every few hundred yards. Sent a messenger into camp and they sent out hot milk for him when we were about 4 miles short of camp. We reached camp, Tatsang (the Falcon's Nest, 16,000 feet), and we were all perishing with cold. K. very bad indeed.

JUNE 5. Kellas appeared better this morning, and said he felt better. It was impossible to stay here, so started him off again on a stretcher. Overtook him a few hours later when he seemed fairly cheerful. Went on over pass to Khamba Dzong. Valley on this side different from any we have been in, and it appears to have more rainfall. Saw a pretty little flowering cistus growing almost a foot high. An hour after I had got to Khamba Dzong a messenger came from up the valley to say that Kellas had died about three hours ago. A little earlier he had sent a message to say that he was going to rest for a couple of hours just short of the pass, and I was not feeling anxious about him. Rode up the valley and brought him in. It is an awful disaster and I feel dreadfully to blame in letting him come from Phari, but I never dreamed then that an ordinary attack of diarrhoea was going to weaken him so much, especially a man of his kind who has climbed much in the Himalaya and who has endured so many hardships. There has been no possibility of stopping anywhere between Phari and here, and I felt quite confident of bringing him here all right. We had planned to send him to Gantok as soon as he was fit to travel....It is quite evident that his hard journeys in Sikkim, when he lost so much weight, weakened him more than he knew.

We buried him in the morning on a spur, with a wonderful view of Himalaya as far as Everest and beyond into the west; Kanchenjunga to the south-west, and to the south-east lie three mountains he has climbed, Chumiomo, Kanchenjhow, and Pawhunri.

JUNE 7. Raeburn has been suffering for several days with much the same as Kellas; he does not get any better and is losing strength. He is really not suited to the strain of this altitude, and I have told H.B. that he is not fit to continue with the expedition at present, but must go to some lower place where he can be properly fed and looked after. Our cooks are unspeakably bad, and in any case cooking up here is very difficult, so the only possible plan is to send him to the Lachen Mission, a three days' journey from here into Sikkim. I shall go with him, as he is quite unfit to travel alone.

JUNE 8. Left Khamba Dzong to-day with Raeburn, whilst the others will go on to Tinki Dzong. Our path lay over a bare open plain to Gira, then it became a steady gentle rise up to the Serpo La (17,000 feet) where we met a most damnable wind blowing straight into our faces. The top was not a regular col, but a hollow saddle about 2 miles wide—bare and stony—and from the further side of it there was a steepish descent through rocks where were large flocks of sheep and goats. We came upon a brilliantly green and purple lake, but unfortunately it was much too cold to stop and look about as I should have liked to have done. There were fine views of Chumiomo on the right and Kanchenjhow on our left, the latter an ugly steep wall of rock and ice. We travelled on through rock and old moraines, expecting to find an encampment of people with yaks, but the place was deserted. As it was now getting past four o'clock we made for a Tibetan tent that we saw about a mile away to our left. This tent was very worn and holey, and we entered through a small opening and down about 4 feet into what was really a dugout. A fire of the usual yak dung occupied the middle, and filled the place with acrid smoke, but the warmth was very comforting, especially to Raeburn who had reached very nearly his limit and was quite done up. I too was about as cold as I have ever been. They gave us hot yak's milk and Tibetan tea—a mixture of tea, salt, soda and rancid butter—for which I am acquiring quite a taste. At all events we got a little warmer, in spite of the wind blowing through the 'roof' and boulders of which the sides of the dugout were made. The people most hospitable and kind. There was a woman busy making butter from ewe's milk in a long cylindrical churn, and it looked pretty hard work. She sang now and again rather pleasing little songs. Another old woman busied herself with the fire and a three-holed oven sort of affair—made of mud and heated by dried sheep's dung. A man or two and an occasional child came in at intervals. Our coolies with their loads showed no signs of turning up, so it was evident that we must spend the night where we were, with no food of our own and no bedding, but later on—out of the darkness—one of our pack ponies with my bedding turned up and we were able to wrap R. up in it. Poor R. cannot eat any of this Tibetan stuff, but the interpreter and I managed to drink the tea. I made the acquaintance of Tsampa, the pounded parched barley which is the main food of these people. Three or four spoonfuls mixed up with a cup of tea make a sort of porridge of quite good flavour—at least so it seemed to me. We also ate some flat and quite tasteless cheese made of ewe's milk, very dry and indigestible, but it filled a gap and we did not starve. We were a large party round the fire that evening and the Tibetans lit a small light in front of a little primitive shrine where one or two of them prayed, or made obeisances. We settled ourselves to lie down, and I curled up in a space about 4 feet long with one side of me against a large boulder. Where the rest of the crowd were I really don't know. There must have been nearly twenty people in the tent and the atmosphere was about as beastly as it could be. But every night has its end, and R. said in the morning that he had been quite warm—thanks to my eiderdown sleeping bag. The Tibetans went out at about four in the morning to milk the goats, sheep and yaks that were in stone wall enclosures just outside. When we looked out, it was to find the ground white with snow and hoar frost. It took some time to get under way, but we travelled slowly on across frozen swamp and river to a small piece of ruined wall marking the boundary of Tibet and Sikkim. This was Girzong, about 15,000 feet, and here we had breakfast in bright sun and a cold wind. Down the valley, one of the upper branches of the Teesta, and here were dwarf rhododendrons, a currant or two, willows, small birches, and after some 8 miles, pines. The flowers were of a much greater variety than those in the Upper Chumbi valley and our journey was even more beautiful, but the clouds were low and showers of rain made photography impossible. At a yak encampment we drank tea with some Tibetans, and I saw a small child of about three years walking about with nothing on, quite unconcerned about the cold. These people were busy weaving cloth when we came up, and many men were spinning wool into yarn.

They seem to be entirely self-supporting, making all their clothes and boots and tents, and living on milk and cheese and 'tsampa'.

I found many little bushy cistus, a red-stemmed rubus, a beautiful little primula just coming into flower, and some lovely clematis—all at about 12,000 feet. I have seen a house-martin at 14,000 feet and a small Fritillary butterfly at 15,000 feet, but I didn't catch it.

JUNE 10. Very good going to-day, and signs of a wetter climate. The trees bigger and draped with long grey lichens, with here and there fine magnolias, sweet-scented syringa, and a Solomon's Seal (?) with lovely white bells. Crossed the Zemu river by a two-spanned bridge with piers made of huge boulders. Then about 13 miles up and down through fine forest to Lachen village, where I at once went to the mission, found two very nice Finnish ladies, and asked them if they would take Raeburn in. They were rather reluctant at first, but then they agreed and quickly made a room ready for him. He arrived a few hours later and in the meantime I was given an excellent tea and eggs. I am afraid it is not a suitable place for R. to spend much time in, so I have sent a messenger to Gangtok—about 50 miles down the valley where there is a hospital—and I have asked them to send up a dandi to carry him down.

I wish I could come down this valley again when we return in September so as to collect seeds, but I doubt if it will be possible. It is a wonderful change to come from the dry bare plains and hills where almost nothing grows, to this streamy valley full of trees and flowers. The air in Tibet is fine and exhilarating, the sky and shadows—especially in the evening—wonderful; but it is not my country and I don't want much of it. There is nothing beautiful in huge snow mountains rising out of a bare plain, and I am not even sure whether there is any real beauty in a snow mountain pure and simple.

JUNE 11. Raeburn better, but food unsuitable for his condition. Three of my fourteen coolies have run away, and six others threaten to stay here unless the man who is now looking after Raeburn comes back to Tibet with me. This forces me to engage a substitute to go to Gangtok with R.; otherwise I should be left without coolies. I have also engaged two ponies in place of the three errant coolies.

JUNE 12. R. much better, so I have no scruple at all in leaving him here.... Travelled back by Tanghu. Passed many droves of yaks carrying loads of madder and split bamboos into Tibet. The little deep sky-blue primula with delicate silvery leaves is now just at its best; it is a most lovely thing, with a delicious scent—not overpowering like the scent of the yellow primulas.

JUNE 13. For a wonder the clouds are not right down and the sun is bright and warm. We were overtaken by the headman of Lachen and invited into one of his yak camps where he entertained us with delicious yak's milk followed by tea. ...My pony gets slower and slower, but the flowers are glorious—the whole valley is scented—so what matters? On the borders of Tibet and Sikkim the people from both sides meet to do trade: the Tibetans bringing wool, barley and brick tea, while the Sikkim people bring madder and split bamboos. I bought a sheep for the coolies for five rupees. I also shot a lark for the skinner (his first opportunity), and a pretty poor job he made of it. Lots of marmots up here; just saw one sitting near his hole....

JUNE 14. Reached top of Serpo La in less than three hours from last night's camp, and got to Khamba Dzong just under six hours. Shot a lark which sings and behaves exactly like our skylark, but seems smaller. There are some stumps of trees here; one living—a sallow of great age—has in it the nest of a pair of magpies.

I have been hearing a lot about the system of government by the Jongpens of this country. It is practically a system of slavery, every house being bound to supply one able-bodied man or woman to work every day for the Jongpen. He naturally grows quickly rich and never spends anything on the place. I am told that about three-quarters of the crops here will find its way into the Jongpen's stores.

JUNE 15. I was awakened this morning by a strong north wind blowing down the gorge and into my tent. Usually the nights and early mornings have been blessedly windless. Left Khamba Dzong and travelled in a westerly direction. Came to Mende, a squalid village on a beautiful plain, but too full of sandflies for me. A clear placid stream watered this plain and there were occasional clumps of willows. Across more sand and grassy plain till we got to Lingga. Here the whole population came out and helped to pitch my tent, and I was serenaded by an old man wearing a hideous mask and two small children who danced and sang very crudely. In the evening it began to rain, soft and gentle. They say that it is the first they have had for more than three months.

I am told that a few people in this part of the country practise cremation, but as a rule they cut corpses in pieces and expose them on the hill-tops for vultures.

JUNE 16. Crossed an absolutely level plain to Tinki Dzong. Passed several shallow lagoons crowded with Brahminy ducks, bar-headed geese, brown-headed gulls, and a bird very similar to a redshank; all with young, and as tame as domestic poultry. Shot a horned lark on the way; they are very common and tame. Tinki Dzong is a picturesque but ruinous mud, brick and stone building, inside a high wall with a large square tower at one end; it covers I should think the best part of 10 acres. I pitched my tent inside the Dzong and afforded a day-long spectacle to the inhabitants. On the whole I liked them, in spite of their dirty skins and their manner of showing pleasure by putting their tongues out. I see a large number of fat pink tongues daily. Called on the Jongpen and found him a spectacled and thoughtful-looking person of about fifty. He wanted to know the value of his wrist watch and his spectacles and other possessions. He has strict orders from Lhassa to help us as much as possible. When I strolled outside the Dzong I heard a banging of stones and found a man pounding madder stems. There was a big rock about 8 × 8 feet, in the middle of which was a 'pothole' about 2 feet across and the same in depth; at the bottom were a lot of madder stems, and these were banged with a large stone weighing I should think 20 lb. I suppose the hole was made by years of banging.

I have seen magpies, barn-owls, Cornish choughs, hoopoes, ravens and redstarts, all nesting in the walls of the Dzong itself. When my coolies caught some of the young geese, at once came a message from the Jongpen asking that they should be let go free. I hear that for a long time there were two lamas here specially devoted to looking after the birds, and it is a particular wish of the Dalai Lama that no creatures should be killed near this place. I never see ill-treatment of animals, and I give a very good mark to the Tibetans on this account.

JUNE 17. Went over the plain on to foothills, and then up a rocky and stony valley. Met a lot of yaks bound for Lhassa and laden with Nepalese paper and boxes containing drinking cups. Camped at the small village of Chushar. Lammergeiers fairly common here; I see one or two on nearly every day's march. Saw the remarkable-looking brown ground-chough. He progresses by a series of apparently top-heavy bounds, at the end of which he turns round to steady himself. At Gyangka I saw two white-tailed eagles, and one let me ride to within 20 yards of him. Brahminy ducks nest among the stones on the hillside.... Saw a pair of white storks.

JUNE 19. Across a howling wilderness of sand; great waves of sand blew 20 feet high in some places. In those parts where the ground was more stony, gorse bushes contrived to grow, making round mounds 1 to 4 feet high. On the lee side (north) of each mound was a long tapering tail of sand showing how it must blow here at times.

Came with suddenness into the main valley of the Arun and followed it up north-west to Chumbab. It was a fine wide valley with a turbid white river, deep here and there, but mostly shallow with sandy rocky banks—very like the Nile below second Cataract. Higher up, this valley widens into miles, with green 'meadows' on either side where great flocks of sheep and goats graze. Chumbab is a ruined collection of huts—apparently an old Chinese barrack—now given over to sheep, goats and donkeys.

JUNE 20. Over an old river terrace to-day, flat and very stony. Remains of old buildings everywhere, and judging by the number of Chortens and piles of stones by the way-side the people here seem to be more religious than those in the country left behind. I am told it is correct to pass these Chortens on the left, i.e. leaving them on the right.

Shot a chough and two rock pigeons (the latter for to-morrow's dinner). A swift with white rump is common in the cliffs here; also kites. Growing in the barley patches is a green herbaceous shrub with a green and purple bell flower: it has a most horrible smell to it and is not unlike an atropos of sorts: said to be poisonous, but in autumn when it dies it is good food for sheep.

JUNE 21. Up the river valley and due north to Shekar Dzong. Here there is a large monastery of 300 or more monks; very squalid village and people even dirtier than else-where. Leaving Shekar Dzong our way lay over two curious sandy ridges with grassy valleys between. From the top of a third ridge the path comes steeply down to the main Arun or Bhong valley. Here the rocks are covered with a very pretty blue thorny Sophera with silvery grey leaves. Many butterflies, *Lycaena* and *Colias*. Crossing the river by a slate bridge of three piers we went up a flat and hummocky valley, the floor of which was well irrigated and cultivated; the round limestone hills reminded me of Cumberland hills. I saw two black-necked cranes just before I got to Tingri Dzong, and here I found the rest of the Expedition, so I have caught them up well. I shall be glad to rest a bit for I have travelled 29 days out of 35 since leaving Darjeeling, and have covered nearly 450 miles. Mallory and Bullock have gone off on a 14 days' reconnaissance of northern approaches to Everest. Everest lies 45 miles south-east, and its rounded dome-like top has the appearance in this light of being a sort of 'core' of the mass. This northern side looks quite unclimbable, it is so terrifically steep.

JUNE 25. (*Tingri Dzong.*) Several sick in camp, two with undoubted enteric. We are living in an old Chinese barrack—the most filthy place for a camp imaginable, filled with the dust of ages and the dirt of every day; the coolies are vile in their habits, and if it were not that many are going away I should insist on getting the whole lot under canvas.

JUNE 26. Howard-Bury has gone off for ten days by himself. Morshead is also away for a few days so I am left alone with the sick. There are several odds and ends to be done, especially in the way of collecting and photography, but it is unfortunate that this is the worst collecting ground we have been in all the way. It is a flat open plain, a great deal of it salt, and there are hardly half a dozen species of plants on it. As for birds, the native authorities don't like our shooting them, which adds to the difficulty of collecting....One of our sick men a bit better, but it looks as if another would die....

People in this part of the country cut a great deal of green turf for winter fuel, and they keep it piled neatly on to the tops of the walls of their houses. But at this season they burn cow and sheep dung which fills the air with an acrid smell. The ravens are extraordinarily tame. I have just seen two women dipping water out of one of the irrigation runnels outside the village and there were three ravens walking round them just as children might do.

JUNE 28. A coolie died during the night. Am determined to move them out of this beastly old building into tents on the river bank. Shall be glad when we move away from here.

JUNE 30. Morshead back from 50 miles up the main river to its head waters. He has brought some *Parnassus* butterflies and a few other things. News from Mallory and Bullock who are getting on to the north ridges of Everest. Nothing definite yet about possible ways of ascent. As usual a strong south wind from midday onwards, but high clouds are drifting from north to south....Found nest with young of a Fringilauda—deep down in a mouse hole.

JULY 4. (*Tingri Dzong. To his Mother.*) ...I am making what collection I can of animals and plants; but of plants there is nothing here growing higher than 6 inches, for we are in the middle of a flat stony salt plain and the flowers are small and insignificant and of no horticultural value. I hope for better things in a few days when I am going over a high pass (18,400 feet) to a valley on the Nepal side, where the climate is presumably a little less dry and the plants may be of a more generous nature. I think Morshead will come with me; he is a very nice fellow and we are lucky indeed in having a very good lot of people. We all get along well together. Up to the present none of us has felt any ill effects from the high altitude, but then one does not run very fast here nor do anything in a great hurry; if you are going uphill you puff and blow, that is all. It is a very great pity that we have no artist among us, for photography is a very poor substitute in landscapes, and in this country the chief beauty lies in the skies and lights. I always wear yellow glasses which improves the colours enormously. These Tibetan people are friendly but very greedy for money. They are indescribably dirty and beyond words ignorant and superstitious. Little Buddhist shrines and images are seen all about the country, and there are monasteries and nunneries full of people who spend their time in prayer and are entirely supported by the active population. It looks like a country that has known better times, for wherever you go are ruins of buildings larger and much more solid than any built in these days. The lamas of the present day seem to have it all their own way, and are opposed to any kind of outside influence which may mean progress....I hope all goes well with you and the garden; how I should like some fresh fruit and vegetables!

JULY 8. (*Tingri Dzong.*) Found two nests this morning: Elwes' shore-lark (with four eggs) and a small lark like a skylark, but nest quite different (with three eggs). Rain and clouds; weather very bad. A note from Mallory to say that he and Bullock climbed a peak of 23,100 feet. Good—but it doesn't make it any easier for me to stay kicking my heels about in this beastly place.

JULY 11. H.B. came back from Kharta this afternoon when I was beginning to think nobody would ever come, so things look much brighter again. We are going to make our next base camp at Kharta, north-east of Everest. H.B. says it is a fine country there with lots of flowers. He and Heron have been eating eggs and stewed wild gooseberries and other good things that make me very envious.

A native woman came to me with much show of secrecy the other day, and produced out of a sack a very aged cat's skin stuffed with hay, and then a freshly killed polecat which of course I jumped at. She wanted a very big price for it (about one rupee!) because the skin is greatly valued by

Tibetans as having the property of restoring turquoises that have lost their colour; it is a pretty fawn-coloured animal with a cream-coloured belly.... A pair of little owls are nesting in the old fort here.

JULY 13. Away from Tingri at last with Morshead. We travelled across the plain south-west to Langkor. There was a biggish stream coming down from Thung-La, which we had to ford as the bridge has been recently washed away. Half the population is now occupied in rebuilding it, and while they work a man blows hideous blasts on a large conch-shell trumpet which is supposed to keep off the rain. There were showers all the afternoon.

JULY 14. At about 16,000 feet I had very good fun catching *Parnassus* butterflies. Later the clouds became thick and it began to hail and sleet heavily; a great pity as there were many little blue poppies and other nice flowers which I should have liked to have looked at.... A long and steady pull of about 12 miles to the top of the pass Thung-La (18,000 feet)—the highest point I have yet reached—and here I found the lovely little gentian amoena. It is not easy to see until you are right over it, when it looks like a little square blue china cup. Saw the Himalayan marmot with its long tail which it whisks sharply from side to side when it is alarmed. It has a curious twittering cry like a bird of prey. One day I heard this cry coming from some rocks and I could not make out what sort of bird it could be—then traced it to this little animal. In a clearing of the clouds we had a fine glimpse of the two tops of Gosainthan—about 30 miles away.... Down the valley again and as it opened out we passed many ruined villages and scattered buildings. This valley has a regular V shape, quite different from that on the other side. Camped at Tulung where the only picturesque thing about the place was the sight of four women weaving cloth in an open space of the village.

JULY 15. On again down the valley. Stopped in a sheltered spot and caught some *Parnassus* butterflies different from those of yesterday. Patches of cultivation, mostly barley; lower down brilliant yellow fields of mustard which they grow for oil to be burnt in their monasteries. From south-west the valley turns south, and thence granite rocks and much more varied vegetation. Came to a small village called Targyeling where we camped in a meadow of turf and flowers. Several birds here that I had not got before, notably a very small *Phylloscopus*.

JULY 16. On to where the valley narrows to a gorge, so that we often had to go high above the river. We got to a village where we hoped to change our yaks, but as there were no yaks available some strong young women carried the loads. Our track was not bad, the sides steep, and we were generally 200 or 300 feet above the river bed. Came to Nyenyam Dzong. The big snow torrent here must certainly come from Gosainthan. The village is the most filthy and stinking place imaginable and the population an unpleasant-looking people—mostly a base kind of Nepalese with a very small minority of Tibetans. No reception from anybody in authority here, so we had difficulty in finding a place for a camp. We eventually pitched tents on a grassy place above the village. As the Jongpen didn't come to see us we thought we would go to see him, but we were told he was too busy to see us. We swallowed this 'insult' quite happily, but unluckily our interpreter let it be known that we had our Lhassa pass with us, which we were keeping for the last emergency in case of difficulty. When the Jongpen heard this, he said he must see the pass and that he could not have anything to do with us unless his nasty little place was mentioned in it. He had the cheek to send one of his men to our camp demanding the pass. He also forbade the people in the place to trade with us, which is an infernal nuisance.

Lots of nice Alpine flowers here: a bushy cistus with

golden flowers the size of a half-crown: two dwarf rhododendrons—one with a very hairy leaf: quantities of edelweiss, a small blue gentian, a big blue aconite, and many other lovely things. The hillsides are bare and stony with some of the biggest boulders I have ever seen. Brushwood fuel is a pleasant change after weeks of yak's dung.

JULY 17. (*Nyenyam Dzong.*) Interviewed the Jongpens. Find there are two of them. They were unpleasant looking people, with long Chinese nails on *all* fingers and dressed in elaborate Chinese silk gowns. The outside of their house was plastered with round cakes of yak's dung and their courtyard was about as filthy as their village. They were fairly civil, but made a great fuss about the name of this place not being mentioned on our Lhassa passport; however we soothed them down, and they said they would help us during our stay, but they requested us not to stay too long! 'We don't want to lose you, but we think you ought to go.'

A lot of talk about the way from here over to the Rongshar valley. We are told there is no way, and that we must return over the Thung-La, which is what we don't want to do.... Saw two of the curious curlew-like birds which are found in Tibet. They harmonize wonderfully with the stones of the river bank and fly rather like a curlew, but more heavily, and utter a very distinct cry. I think they are ibis-bills. They have eggs or young on an island in the torrent, but unfortunately out of reach.

JULY 18. Across main river with Morshead and down the valley for a few miles. We got into some very tangled mountains when clouds came down, so we turned back. Had fine sunshine in the afternoon for drying skins. A beastly village dog has just stolen the marmot's skull which I was drying at my tent door.

The authorities have tried to prevent us going over to the Rongshar valley, but they have had to give in at last. They have not helped us in the least, and have charged exorbitant prices for the foodstuff we have bought. Shall leave Nyenyam without any regrets. We have decided to send a message to the headman, who will pass it on to the Jongpens, that we came here with valuable gifts for them all, but as they have done nothing for us we are going away without presenting them. Hope this will annoy them.

Arranged for coolies to carry our things over to Lapche to-morrow.

JULY 20. (*Nyenyam.*) An exquisite primula grows here. It has three to six bells on each stem, and every bell is the size of a lady's thimble—of a deep blue colour and lined inside with frosted silver.

Left Nyenyam with a horde of natives, male and female, and one yak. Easy steep track to the foot of a small glacier. Went up this in the tracks of some natives and found it was a good deal crevassed near the top. A round-topped col (17,800 feet) between two small peaks—no view. The whole of this west side of the mountain slope is sacred as having been the haunt of the Saint Mila Respa, and is known all over Tibet as Lapche Kang. Hoped to reach the village of Lapche to-day, but our Tibetan coolies came along very slowly so we camped in a rough spot by a river while the natives found shelter in some caves close by. Our fuel was green dwarf rhododendrons, and our cook was quite unable to deal with it, which made me long for some of my good Dayaks who can make a blaze out of the wettest stuff. Before starting next morning I was amused at the way in which the Tibetans apportioned the loads. Each man or woman takes off one of the garters with which they tie up the cloth tops of their long boots, and this they give to the headman; he shuffles them together and then puts one on each load; the one whose garter is on the load carries it.

JULY 22. Down to Lapche village. This valley much wetter than the one we came from. How wonderful it must have been a month ago when the weather was fine and the primulas were out. There are still quite a number of the large yellow primula, with drooping heads and stems nearly 3 feet long. The houses of this village have pent roofs—slabs of wood weighted with stone like those in Sikkim. I do not like to shoot the birds here in this very holy place, for the good Buddhist objects strongly to taking life, and I find my Lepcha collector—who loves shooting birds—doing most profound obeisance to the images in the little shrine here. We camped on a level terrace beside the famous Lapche temple. It is a square plain building with a Chinese-like roof surmounted by a bright copper ornament. An old Lama and an old woman seem to act as caretakers, and nobody would guess that it is one of the most famous places in Tibet. Buddhists from everywhere make pilgrimages here. Mila Respa, poet and saint, was—nobody can tell me how many years ago—a Tibetan incarnation of Buddha, who came and lived in caves and holes about this mountain until he was taken to heaven like Elijah. He seems to have had a certain sense of humour, for one day he was walking with one of his disciples and found an old cow's horn lying in the path. The disciple said it was no use and passed on, not seeing that Mila Respa had picked it up and put it under his cloak. Shortly afterwards a violent storm came on, caused by the saint, who thereupon took the old cow's horn from under his cloak and got inside it for shelter. 'Now you see', he said, 'nothing in the world is useless.'*

JULY 24. Up valley north to-day. Crossed a river to the left bank by a curious natural bridge of boulders, then steeply into a side valley where we camped at 15,200 feet. Here was the upper limit of the small rhododendrons. Found a pretty

* In *Mount Everest, the Reconnaissance*, 1921, published by Arnold 1922, A.F.R. contributes a delightful account of this excursion he and Major Morshead made to Lapche Kang, where no European had before penetrated. He also, in the same book, devotes a chapter to 'Notes on Natural History'.

new white primula—4 to 8 inches high—with a delicate primrose scent. We hope to cross over to the Rongshar to-morrow. We had thought that Lapche was in the Rongshar valley but the maps are all at fault.

JULY 25. Nearly three hours to the top of a pass, and down a few hundred feet to a large and very broken-up glacier covered with big moraine. Ice only showing here and there where big lakes—like small craters—make little lakes. Shot two rose finches at 16,500 feet. Rather complicated track down the slope, as it went steeply up for several hundred feet over two steep spurs which is quite unexpected in descending from a mountain. High over our heads, topping the clouds, was the summit of a most beautiful peak—Gauri-Sankar. Down through flowers to Rongshar, lying on a rocky shelf high above the river. Little patches of cultivation, such as buckwheat and poor-looking potatoes; also big larches, the first I have seen in Tibet (but this is not really Tibet). We wanted to buy a goat here as we are short of meat. We were told that we might certainly buy one but we must on no account kill it in this valley, as it is sacred and no life may be taken here. They offered us yak meat bought at Tingri the other day, but it smelt too strongly for us.

Took a photograph of Gauri-Sankar. It is a most beautiful peak, with an amazing knife edge of ice-rock running for thousands of feet up to nearly the summit.

JULY 27. Up the Rongshar valley, which meant a preliminary descent of 1800 feet to the level of the river. Here it is a great torrent in a winding and horrible gorge—to me very depressing. A long and tiring march to Tazang, where the only available place for our tents was about the size of a billiard table with a big tilt to it. I slept in a Tibetan tent full of holes. Any quantity of wild gooseberries here; they are just in the stage for stewing so we shall experiment on them to-morrow.

JULY 28. Made a short march to an open space below a glacier where there was a stone hut used by herdsmen and plenty of yak dung lying about ready for fuel. I was glad we had to stop here as I got two birds new to me: an accentor, and a very dark brown, almost black, wren. Also two of a new mouse, darker in colour and with larger ears than those I got at Tingri. Have acquired an abominable cold in the head which makes me stupider than usual.

JULY 29. Up to the Phüse La (The Pass of Small Rats), then through dirty old moraine stuff to Keprak, a dirty village in the middle of a stony wilderness. No cultivation, and one wonders why anybody dreams of living here. Bought a very old sheep. It is better than none, for we have been meatless now for a few days.

JULY 30. Down the Keprak valley, and then over a broad spur into another equally dreary valley to a pass at about 17,000 feet. The surroundings all very ugly, and the rain and hail prevent our seeing the splendid views of Everest and the neighbouring peaks. Near the village of Chobu there was a group of thirty or more willows of immense girth; they must be of great age. Gazelles looked on at our tent-pitching quite unafraid. This is another very holy valley where nothing may be killed. Most of the people seem to be monks.

JULY 31. Down a wide valley, and then southwards to the village of Rebu. Bitterly cold, and fresh snow on hills quite low down. This makes me wonder what September will be like.

Reached the Expedition's camp at Kharta and there found Howard-Bury.

AUGUST 3. (*Kharta.*) Some of us have been to tea with the headman of a neighbouring village. There was a tent pitched in his garden and we sat on the usual stool covered with carpets. We were served first with peas in their pods and rose hips—both raw; then various dishes of macaroni, chopped meat, funguses and other vegetables, which we ate as well as we could with chopsticks. Afterwards our host produced a wonderful three-stringed instrument of the banjo kind, made of wood, with two sounding drums and a long head curved into the fashion of a horse's head. He played and sang to us not unmusically.

AUGUST 5. Howard-Bury has gone to join Bullock and Mallory up the valley where they are looking for the way to the east side of Everest. I went with him for about 3 miles, then I struck off to the left and went up to the Sumja La. I got down into the valley beyond and found a perfect mountain stream connecting some beautiful little lakes with bays and rocky points: clear water and stony sides exactly like many of the lakes or tarns in Sutherland. Except tadpoles, there was not a living thing to be seen in any of them. If only there were I would come and camp here for a month. Hardly any birds or butterflies, but growing about the lakes was a fine purple iris with a creamy splash on it. A claret-coloured poppy, now going to seed, grew in the dry places.

AUGUST 12. Weather still bad. Monsoon clouds and rain penetrate through the gorges to this side of the main range, but when it abates we shall move up nearer to the mountain and make what kind of assault on it we can.

Tibet is a fine spacious country—plenty of sky and colour, and all on the grand scale. I am glad to have seen it and the Himalaya, but it hasn't stirred me as some other countries have.

AUGUST 20. Mallory and Bullock back from their high camp up this valley. They report snow very soft, and impossible to do any high climbing in these conditions.

Went with H.B. down the gorges of the Arun as far as we

could go. It was not many miles in a straight line from here, but it was quite different country and totally different climate.

AUGUST 23–25. Away from Kharta with H.B. At the head of a valley, alongside a huge moraine-covered glacier, we saw some small black rats playing about the stones. Of course my boy was miles behind—for which he got properly cursed—and I had no gun to shoot with. On over a pass, where is a stone wall of defence built by the Tibetans against the Nepalese, and then we stopped near a lake and waited for the coolies to arrive. This lake was called the Lake of the Bead. Long ago an old woman dropped a bead into it and it was finally found at a place in Nepal. Now the people of that place come once a year to make festival by the lake.

Down the valley brought us into luxurious vegetation, beginning with willows and going on to large juniper trees. Saw the Himalayan crested tit: very familiar note, easily detected. Found some very big red currants, bitter but quite good, and we had them for dinner. The Kama is just visible below us, and but for the clouds being low we should get a splendid view. At 11,000 feet I saw a langur monkey on the path. Could not make out what it was until he climbed up a big tree to look at me. Descended steeply to the edge of the river torrent which comes from the glaciers and eastern slopes of Makalu and Everest. All the trees were draped with long beards of lichen, and there were giant birches, sycamores and alders, together with a number of flowers new to me. Everything dripped with water, and for a long way we walked on narrow logs laid on mud and water.

Crossed the river, and came to a large clearing of the forest covered with long grass and simply swarming with leeches, which I did not believe came so high as this. Camped at the upper part of this clearing and spent most of our time removing the leeches from our tents and our persons. A large pair of scissors was very effectual.

Down to a junction of the Kama with the Arun, and then our path got hotter and steamier every hour. The sun was bright and strong and I saw many butterflies. There was a curious species of rubus which looked as if the stem and branches had been whitewashed. Into the Arun valley, and then through cleared grassy slopes to a sort of terrace with some cultivation. At the little village of Lungdo the people were greatly interested in us as no Europeans had come this way before. They brought us pumpkins and excellent little cucumbers, as well as a cake of delicious honey toffee.... Started back in broiling sun, and after some few hours' walking I felt about as 'cooked' as I have done for years. When I came to a cave under a huge boulder at about 4 p.m. I made up my mind to spend the night there, but unfortunately there was no water near so I had to drag along. As I got higher up I felt less exhausted, but I was extremely glad to get into camp after thirteen hours of very hard going.

Next day was through beautiful woods following the course of a very pretty mountain stream. Astonished to find leeches as plentiful here at 12,000 feet as down below. It was an easy ascent to the Popti La pass, but clouds were down as usual, so no view. This is the Tibet and Nepal frontier, and it is marked by a low wall of stones and turf, hardly enough for defence. On the Nepalese side is a very pretty lake with one or two rocky islands on it—for all the world like many of the lochs in Sutherland. There seemed to be very minute fish in it, but I could not make out for certain.

AUGUST 26. Last night and to-night we have had a splendid camp fire of spruce, juniper, and rhododendron outside our tent after dinner. The sad thing is the enormous number of moths that die in the fire instead of in my killing-bottle.

Many Tibetans come over to this valley to collect the tubers of the wild arum—the kind called cobra head in India. They foment it for a few days before it is fit to eat, then eat what to my mind would be an unpalatable bread.

Saw a bat, the first I have seen in Tibet, but did not shoot it.

AUGUST 27. Up valley north; fine sunshine but mountains hidden. Shot a long-eared mouse hare—the high mountain kind. Shot and lost a moraine wren which fell under a big boulder; this was most annoying as they are very rare.

AUGUST 29. Climbed to the top of the ridge west of camp, and followed this ridge to a small peak, just above the Chog La, where I overtook H.B. who had shot a Tibetan snow partridge. Down loose screes to the Chog La where I got one out of a flock of beautiful purple-winged birds (Hodgson's grandala). Came upon three stoats, very pretty fawn-coloured animals with white throat and yellow waistcoat; shot the one which stopped to peer at me from the shelter of a dwarf rhododendron.

Down to Kharta and into quite different climate. Dust on the path and signs of autumn in the barley fields; willow trees all turning yellow.

AUGUST 31. (*Kharta.*) Bullock and Mallory leave for Kharta valley to prepare for attempt on Everest. Weather still bad and they can do nothing up there until a change to cold.

SEPTEMBER 5–7. Left Kharta with H.B. and Raeburn (who has rejoined us from Darjeeling) and went up the valley. Camped beside river at 14,800 feet and then up a steep pitch over the end of a side glacier valley. We arrived at advanced base camp, 17,350 feet, and found Mallory and Bullock. Here the valley divides into two and is much more open. Many Lammergeiers and kites. A white andromeda in flower on the slopes. Wandered about collecting seeds, and climbed a long stony rounded ridge to 19,500 feet....H.B. has seen a red fox at about 18,000 feet.

SEPTEMBER 9–16. Snow showers day and night. Quite impossible to do anything up high. The glaciers are about 3 feet deep in soft snow. Deadly monotonous up here, and the wonder is that we don't all bite each other's heads off, but we don't. For the first time since June 7 all eight members of the Expedition are together. We live in two messes as one tent will not hold us all. I have managed to do a certain amount of collecting in spite of the snow, and have several new birds and two new mice. H.B. has seen a grey wolf and a hare at about 19,000 feet. I have seen painted snipe several times at 17,000 feet, but didn't shoot one; also a hoopoe, dippers, and black-eared kites.

Curious saussureas here: large composites packed with stuff like cotton-wool, and such is the warmth inside that on the coldest day a bumble bee will often come buzzing out from one of these plants.

SEPTEMBER 17. H.B. and others up a ridge to climb Kama Changri (about 21,300 feet, to south). Mallory and Morshead were quite exhausted at the top of this climb, which does not look well for our Everest attempt.

The weather cleared finely yesterday, looking like the end of the monsoon, but this morning broke badly with low clouds and snow—just like another puff of monsoon. This delay is very serious, as the days are getting much colder and the nights too. Also with all this new snow high up there is very little evaporation or hardening of the crust. The long snow trudge and climb of 3000 feet from the next camp to the col of Everest is quite impossible except in good conditions. What I most of all fear for the climbers is cold and frost bite, and I confess I dislike the prospect of going up to the second camp (23,000 feet) from which they will make their two or three days' attack on the mountain. Even down here it is most unpleasantly cold at night, and disagreeably chilly all the time. There is no question but that the right time to attempt the climb is in May–June, before the monsoon.

SEPTEMBER 21. Up to 20,000 feet camp to-day....Nice walk up valley; stones and occasional patches of grass, but the last part all stones and going bad. Over a moraine we came to a glorious view of Everest and its magnificent south-east ridges. I took it very slowly and came up in seven hours. The camp is on a stony shelf, a few hundred feet above glacier and immediately below the sharply cut-off edge of ice-cap from hill above. It looks like the edge of an ice-cap in Greenland or Spitsbergen, not a bit Alpine. Mt E. about 5 or 6 miles away up glacier, and very fine at sunset; it bears a little south of west. The night very cold, but we are open to the sunrise here and so had breakfast outside our tents and were quite warm in the sun. Went up a stone snow-covered hill to take some photographs. Mt E. from this side certainly magnificent, and makes up for its dullness from the north. Makalu very fine, and Kanchenjunga an enormous mass rising above a sea of cloud with the pointed peak of Jannu near it....I stayed up here quite warm in the sun.

SEPTEMBER 22. Six of us were away from camp by 4.30a.m. Down the glacier, and there we put ropes on to the coolies to keep them together....Then began a long trudge up glacier. The moon, nearing her last quarter, gave plenty of light at first, then dawn about 5.30. Crisp snow, and a gentle ascent for about 2 miles; all quite easy. Towards nine o'clock the sun began to get very powerful, the snow softer and the ascent steeper. About 21,500 feet we saw curious tracks in the snow, very like those of a small man walking fast and easily. Coolies assured us that they were the tracks of one of the wild men, a small people covered with hair. My notion is that they are the tracks of a wolf, loping along at just the sort of pace to make its four feet give only two tracks. At about the same altitude or a little lower, I saw the tracks of fox and hare. The only birds seen were ravens, a montifringilla, and a hawk about the size of a sparrowhawk but very pale underneath. Going now became difficult, and the heat on the glacier—hemmed in by mountain spurs on both sides—was almost intolerable. The difficulty of lifting myself up for a few steps at a time was as much as I could manage, so how the coolies carried their 30 lb. loads I cannot think. Somehow we all did it, and arrived at the col we were making for (22,350 feet) by twelve o'clock. I was nearly the last, but going well enough. In the matter of colour we were all about equally blue. A nasty wind blowing, so we pitched camp in a slightly sheltered depression on the col without losing any time. I was glad to find that I suffered less from cold than some of the others, in spite of my wearing only one pair of socks and comparatively light-weight boots rather loosely laced with puttees above.

From this col is a steepish descent of about 1000 feet to a glacier, leading in about one and a half miles to the north col of Mt E. From here it ought to be quite possible to go for some thousands of feet up the north-east ridge of Everest. Apparently this is the only practicable way to the summit that we have seen. It is not going to be at all an easy job to get up the col itself and I don't think there is any possibility of taking coolies up it to make a camp from whence they can climb further, but if our people can get on to the col themselves they will prove that this is a practicable way of approach to Mt E. and very likely the only one.

We crawled into our Meade tents, two in each, and though they are badly designed tents, any port in a storm....We had brought no boy with us so had to do what cooking or melting of the snow was required ourselves. We managed a sort of meal of half a cup of consommé each, a scrap of ham and biscuits, and about half a cup of tea which froze to the snow at once if you did not pick it up. Turned in about 5.30 and spent the most beastly night. I had snatches of sleep, but I don't think H.B. in the same tent got any at all. The ground under us was solid frozen snow, and as the night went on the lumps and bumps underneath seemed to grow into Alps and Himalaya. The temperature inside was well below freezing, our boots were solid, and outside it went down to zero Fahrenheit. Thawed a little about 6.30 next morning when the sun touched our camping place, then managed another meal with a little tea and sardines frozen solid in their tins. All of us very blue in colour. Took a few photographs, but fingers were frozen by touching the brass screws of the camera.

It was decided that Mallory, Bullock and Wheeler, with ten coolies, should go on and make an attempt to get on to the north col of Everest, and that we others should go back to our 20,000 feet camp and wait for their return. We were all very stupid and muddle-headed and so took rather a time to get off. I do hope they will all come back none the worse for it the day after to-morrow....Weather very thick and snow falling....Got back as fast as we could to our camp which seemed like the height of civilization. Tea extraordinarily good.

SEPTEMBER 24. Quite tired enough after the last two days, and glad to loaf about and feel warm stones under foot instead of frozen snow. A strong wind is blowing off the crests and ridges of Everest, which looks bad for the climbing party.

SEPTEMBER 25. About 4 p.m. the mountain party returned, and greatly to my relief none of them was any the worse for it. Wheeler had suffered more than the others and was yesterday evidently at the first stage of frostbite, but he is all right now. From 'windy col' they had gone down about 1200 feet on to the glacier we saw; then up it for about a mile, and camped at 22,000 feet. The next morning, with three coolies, they kicked their way up 1500 feet of very steep snow, and in two and a half hours got to the col between north peak and Everest itself. There they were met with a furious north-west gale. They say the ridge up from col north-east looked quite practicable for several thousand feet, but there appeared to be no likely place for a camp higher up. Anyhow they have finished what the expedition for this year set out to do, i.e. to reconnoitre Everest and find a possible way up. I have never thought that it could be climbed, for physical reasons, but if it is again attempted the climbing party will have to be in the very best condition, and fresh, and they must have perfectly still weather for a number of days. They must also have a man to cook for them who is accustomed to high altitudes; none of this 'heating up' of tins, and melting of snow by members of the expedition as we have done. Mallory says he felt equal to going up another 2000 feet, but the others I think were both at the end of their tether. We spent a relieved and cheerful evening in my 40 lb. tent; a tight fit, but it made for a little warmth....

SEPTEMBER 26. H.B., Wheeler and I, went with coolies across the Kharta Glacier to the Karpo La, about 2 miles to the south-west of camp. It was easy enough for anybody accustomed to climbing, but a great deal of loose stuff lying about caused the coolies to make very heavy weather of it. I let them down the only difficult pitch on the rope. Thence down a glacier, off on to moraine and boulders, and to the bed of a small stream running parallel to the huge glacier that comes from below the cliffs of the east side of Everest. Our stream dived into the glacier and we went along an old moraine shelf which was quite good going. Eventually we came out on to a wide grassy plateau where was an old yak-grazing encampment, and here we camped with great contentment; turf instead of stones under our feet, and breathing with comfort; night temperature 18° Fahrenheit instead of the zero we have been having.

SEPTEMBER 27. Woke to one of the most perfect mountain days it would be possible to imagine. Breakfast in quite hot sun outside the tent, and then away collecting seeds. H.B. off on a tremendous walk and climb to a col (21,000 feet) between Makalu and Everest, whence he looked down into Nepal and saw views that no man ever saw before; a very notable achievement. He was rather done when he returned at 5 p.m., and as he is twice as good a walker as I am it would have been useless my attempting to go with him, though I should have much liked to have done so.

SEPTEMBER 28. Disappointing change in the weather to-day; the lower valley filled with clouds which roll up and hide everything from us. About a mile down the Kama valley we passed the end of the great Everest glacier and came to the amazing sight of a huge glacier coming out of the middle of Makalu. It pushed its way right across the Kama valley in such a manner that the river coming out of the Everest glacier had to go underneath the Makalu glacier, which it entered by a large cave in the ice. The M. glacier must have retreated a little recently, for there are obvious signs of its having butted up against the cliff on the left (north) side of the valley; now there is just room to pass over boulders and stuff between the cliff and the snout of the glacier. Crossed a small stream, and then on to a little grassy meadow where I lay on a bed of cistus and ate my lunch. It was a most heavenly place, and I watched the tops of Makalu appearing and disappearing in the drifting clouds. Then up a steep bank covered with deep blue primulas and juniper trees to a small level terrace where we camped on a lawn blue with gentians. Next day we travelled down the valley over low hills and across pretty streams, past one lovely lake, and among great fields of iris and gentians. The autumn colours of the berberis and mountain ash are gorgeous. We pitched our tent amongst some enormous boulders. The coolies saved themselves the trouble of pitching theirs by finding shelters under the boulders, and they soon had good fires going of wet rhododendrons.

SEPTEMBER 30. Up about 1000 feet to a more or less level terrace where I found plants of the Giant Rhubarb, now no longer white but red and with seeds ripe inside. A gentle slope up to the Shao La pass and so back to Kharta, where all the barley has been cut and carried since we left a month ago. Real autumn; the trees and bushes lovely—full of gold and red.

More than once lately I have heard at night the cry of migrating waders, curlew being unmistakable.

OCTOBER 2. (*Kharta. To M.M.*) ...If you often write letters that give as much pleasure as this, you deserve more than a crown of glory. It reached me about a fortnight ago, in a high-up camp where we were snowed up for days, and it did me a world of good. I like to be told to shut up when I talk rot about my old age, and I very much like to be liked, so if this is the kind of pick-axe you threatened to use on me, please go on with it and you may make something of me in time! Talking about age is, I admit, a silly affectation; for as a matter of fact I don't feel old a bit, and I get as much joy out of life as ever I did, and so may it continue....Our time here has been extraordinarily interesting, but rather a picnic compared with New Guinea. I don't know that I ever want to come back. My chief delight has lain more in the indescribable beauty of the flowers and the sight of these mountains, than in the conquest of Everest itself....

OCTOBER 5. Left Kharta and away up the Arun valley to the Bhong-Chu, which we forded without difficulty; camped at Lumeh. Next morning I saw one of the most astonishing bridges I have ever come across. It was three twisted strands of hide, stretched across 30 yards of torrent from short posts on either side. Over these was a sort of yoke of wood, free-running, and pulled by ropes from one side to the other. From the ends of this yoke hung long leather straps, to which were tied the load or the owners thereof. We sent all our things over in this manner and to my surprise there was nothing lost. The pulling was mostly done by women.

Camped among willows at Khar Khung.

OCTOBER 6–10. Up the valley to Lashar, and then to Tinki. The flat plain near Tinki, over which I galloped in June, is now flooded, so we had rather a roundabout way to Lingga. Enormous quantities of wild fowl here; garganeys, bar-headed geese, sheldrake, gadwall, pochard, Temminck's stint, and redshanks. Bitter cold south wind. People busy threshing barley.

OCTOBER 11–13. (*Khamba Dzong.*) Fixed up stone over Kellas' grave, and inscribed in Tibetan and English characters his initials and the date of his death. Away and to the top of the pass where Kellas died, and camped at Tatsang. One of the coldest nights I ever spent; temperature down to zero—just under outer fly of tent; boots and everything frozen solid.

OCTOBER 14. Whole country white with snow. Mountains wonderfully clear. Up and down all day, altogether about 25 miles. Much too far for coolies and animals as snow is very heavy in places. H.B. and I huddle together over a fire in a small leaky Tibetan tent waiting for transport to arrive. When it does, everything is so frozen that it is hard work to knock tent pegs into the ground.

OCTOBER 15. Many coolies snow-blind from yesterday; they put their goggles on too late. The Tibetans tie their pigtails across their faces below the eyes and do not seem to suffer at all.

On to Phari. People everywhere busy harvesting their barley; it has practically no ear and is used mostly for animal forage. Very pleasant to get into a house again after all these months. Contrive to keep warm by a wood fire, but it is still bitterly cold outside.

OCTOBER 17. Great delight to get away from vile Tibetan plain and down into valley among trees again. It seems that autumn is already over and winter has begun. Collected seeds of rhododendrons which I had marked on my way up. The berries of the mountain ash here are pure white.... Went a few miles up the Kambu valley and got seeds of a pretty *Enkianthus* seen last May.

OCTOBER 20. Snowing hard all night and morning. A great disappointment as I had hoped to collect a lot of seeds going over the Jelep Pass, and take photographs. Everything in a mess with melting snow but it is getting a bit warmer as we go down hill. Quite hot at Sedongchen. A big drop to-day from over 12,000 feet at Gnatong to 3000 feet at Rongli. Here are little green oranges, very good, and it feels quite tropical.

OCTOBER 22–25. Rongli to Pedong. A tiring day for our animals, and my pony completely gave out about 3 miles from Pedong, so I had a hot walk up. Great numbers of large black and yellow spiders everywhere; many little green parakeets....At Kalimpong my pony collapsed, so I had to hire another. Splendid view of Teesta valley, with tops of Kanchenjunga and others showing. Peaceful afternoon in the sun—temperature just about perfect now.

Downhill through rice fields and forest to Teesta Bridge, and so back to Darjeeling. Got rooms at the Club, and had luncheon with admirable beer! Busy with all our goods and collections.

NOVEMBER 7. (*Delhi. To M.M.*) ...I will inflict another scribble on you, and this from a more civilized place than the last, though I am not sure that I wouldn't sooner be there than here; for now I am playing at being a globe-trotter for a week or two and it is not a great success, as I am a wretched sightseer and generally avoid show places. I once spent a week in Venice without going inside St Mark's. Perhaps you will think me an awful Philistine and perhaps

I am, but in any case wandering about Indian cities alone is a deadly amusement. I would give anything for a congenial companion. Indian buildings, so far as I have seen them, are fearfully overrated and not to be compared with the Moorish buildings in Granada or Seville; their arts and crafts are beneath contempt. Perhaps I must not damn them utterly as I have not yet seen the Taj Mahal at Agra. After that I am going to make a bee-line for Peshawar and the Khyber Pass, which attracts me much more than all the tombs and ruined palaces in India. You will think I am in a nasty discontented frame of mind, seeing no beauty in things that many people would give their ears to see, but I am really in rather a nice kind of mood—only hating the restrictions of houses and cities. No doubt I shall get used to them soon, but I am sure I am right about Indian art and architecture.

DECEMBER 12. (*King's College.*) I have missed Founder's Day at King's—a thing I have been looking forward to for months. However, I spent the 8th at Eton for Founder's Day, and Monty James put me up and was as dear a man as ever. Wonderful service in Chapel, with King's and Eton choirs and a part of the London Symphony Orchestra. Dinner in Hall; afterwards a quaint Latin play by small boys, such a good evening. A few days later I went to King's and 'hung up my hat' as they say, for the first time for twenty-five years. Since I ceased to be an undergraduate in '96 I have lived either in lodgings or in tents or ships, and have had no place to call my own; so this is quite a pleasing experience.

DECEMBER 20. Great 'Everest' meeting at Queen's Hall. Was supposed to have spoken, and had screwed myself up to required pitch, but H.B. was followed by Mallory so that there was no time for any more. The D. of York was beginning to yawn, so the meeting broke up and my quips were wasted. Have definitely decided not to go on next year's expedition—wisely I feel sure.

IX

TO COLOMBIA

MAY 1922. (*King's College. To his Mother.*) What a lovely country England can be when it likes. Here the trees and grass are about at their best. Lilacs are just bursting, but there is not yet a sign of yellow on the laburnums. What do you think of kingfishers and goldfinches nesting in a town like this? I have seen both this afternoon. Nightingales are still very common, but they are not quite so near to the town as they were formerly, and I have to listen very intently to hear them from my window at night.

About this time A.F.R. still hoped to raise enough money to equip another expedition to New Guinea, but by the summer of 1923 the difficulties of securing funds became very great.

I fear I must give up New Guinea—at any rate for the present. Tom Longstaff* has put into my head the idea of visiting the Sierra Nevada range of mountains in Colombia. This range is quite distinct from the Andes, and it lies to the extreme north-east of Colombia. It is a range very little known to Europeans so I think I might go and have a look at it with a view to possible further venture.

In the autumn of this year A.F.R. and I became engaged. We were married in King's Chapel on November 15, and ten days later sailed from England for the port of Santa Marta, Colombia.

On December 12 in lat. 13° N., I saw Venus with the naked eye at 2.10 p.m., half an hour beyond its meridian; bright sun and very clear. Next day we came into Santa Marta Bay and set about to get various information about the S. Marta mountains.

* It was Dr Vaughan Cornish who first called Dr T. G. Longstaff's attention to this piece of exploration.

DECEMBER 22–29. (*Santa Marta.*) It is ten or fourteen days along the coast to Rio Hacha, our point for getting to the Sierra Nevada. We turned down the offer of a dugout canoe to take us along this hundred miles of dangerous coast, and on the 22nd we decided to get a passage on board a small schooner of about 30 tons. There were two little dog kennels aft—by the wheel, measuring $5\frac{1}{2} \times 2 \times 1\frac{1}{2}$ feet, and in these we unrolled our valises and tried to lie down; extreme discomfort. Got out of bay into rough water and my poor M. absolutely prostrated. Very heavy seas. I too began to be sick and altogether it was the most poisonous voyage I have been in. We were told that we should be in at daylight on the 24th, but it was nearly midnight before we anchored off Rio Hacha. Landed on the 25th in a large dugout canoe with Customs officials who examined our baggage for contraband rum and tobacco. Took the only available room in a ramshackle 'club' affair, and then went out to interview various people about mules. This is a scrubby little town, one-storey houses, rectangular streets and the ground sand. Water is fetched about 3 miles in barrels. A small boy sits on a donkey, and the donkey drags one barrel by a rope while another barrel is attached to a rope round the middle of the boy....Spent two days trying to get mules. These Colombians are dishonest thieves, they lie to you in the most exasperating way, ask absurd prices—in fact they are the very deuce.

On December 29 we got three mules after much haggling, at 72, 90 and 60 dollars. Spent that night getting our gear in order. Started up at midnight and by 1.30 a.m. were on our road and mighty glad to leave Rio Hacha. For several miles we went through dry thorny scrub, then got down to seashore which we followed for the rest of the day. Crossed two rivers in canoes with mules swimming. Country flat as a board, alternately dry and mangrove swamp. Many pelicans along the sea, and a few cranes in the swamps, otherwise very little life. Mountains about 10 miles inland on our left, sea

A. F. R. IN 1924

on our right. Water always at high tide so much of the time is spent splashing through it, and we have great trouble in getting the mules along. Very hot and tiring day, but M. as happy as I am, so all is well. Reached Dibulla after 13½ hours' travelling. It was a small village of roughly a thousand people and we were lodged in a nice empty house where half the village came to look at us eating. We had a splendid bathe in the river, but I was a bit nervous about crocodiles.

JANUARY 2, 1924. Started off with oxen, and after about a quarter of a mile one of these beasts took fright, chucked its load, scared the others so that they followed suit, and loads and saddles got flung about and smashed in every direction. We had to return, much to the amusement of the village, and it took us a whole day to get together again. When we did get off we had difficulty in crossing a river. Two mules upset, and both I and our Jamaican guide were thrown into the water.... Went first along shore and then by extensive coconut plantations; later almost due south inland. Sandflies and horseflies very troublesome. Got into edge of forest and began to rise gently. Well-made old Indian road—but lacking in some places. Camped in forest at 700 feet.

Up and down steep ridges, with dense forest—like Ruwenzori. Very little life. A few flowering trees and here and there open patches of old cultivation. Our path is quite good in the places where it is paved with large rounded stones from stream beds. These paved ways probably date from before the Conquest. Parts are so steep that the oxen have great difficulty in getting up, and one beast broke down entirely and we had to leave it with its load. Went over a ridge at 3500 feet into a valley (going south-south-west) to Pueblo Viejo. This is a village of about twenty huts, wattle and daub, with grass thatch. People very negro in appearance—some almost pure. We are at edge of forest on cleared hill slopes. These slopes above the village are almost entirely denuded of trees by long cutting for firewood, and the grass harbours innumerable small ticks. The nasty little pests are almost invisible and cause intense irritation to the skin. M. and I spend our evenings de-ticking one another before going to bed. A bathe in the clear cold water of the river is a temporary relief to the irritation.

Have seen a few humming birds, all of one species, but on the whole birds are few and mammals almost absent. Several Indians from the villages higher up the valleys have been visiting us. They have never before seen a white woman and believe that Mary is my son. Perhaps this is because of her long hair, for the Indian boys wear their hair long; or perhaps only because of her breeches; at all events they are fully persuaded that we are father and son.

Left Pueblo Viejo with a train of six bulls, and went up a very steep slope for about 1000 feet south-west, and so over into another valley of the Rio Ancho for about 4 miles. Again part of our road was paved with large slabs of stone—evidently of great age. Reached San Miguel, 5400 feet—an entirely Indian village. They grow here small quantities of the wild canna and eat its roots; we tried it and found it excellent. San Miguel has about 60 huts, mostly round beehive shape, wattle and daub and grass thatched; a few made of latticed reeds. It owns a 'strangers' house', and a square church used once a year by Catholic priests on their visits. The entrance to the village is made through a sort of lychgate with heavy wooden doors. We are told that when a new house is to be built the roof is made first, put upon the ground, and then the place for the walls plotted out. These natives were quite friendly and are allied to the Arhuaco Indians. They are a small tribe confined to this mountain range and they call themselves Cugúi. They are not much to look at and have a very poor physique. The men grow their hair long, have little on their faces, and just one or two have beards. The tallest man we have yet seen cannot have measured over 5 feet 5 inches. They are distinctly puny and their skin is a pale mahogany obscured by dirt. They are ceaselessly occupied in taking coca from a small gourd held in the left hand. The dry coca powder is mixed with lime, and with a long thin stick dipped into the gourd they bring out a little of the mixture and suck it off. The stick is then dried by being rubbed on to the open end of the gourd, thus forming in the course of months a lump of dried stuff as big as the hollow bulb of the gourd itself. This habit of coca taking has produced a drugged sleepy look in their faces, and no doubt accounts for their wretched appearance. Immense pains are taken in the cultivation of the coca plants, which are grown on terraces built on the hillsides and easily irrigated in the dry season. Most of the hillsides are cleared of forest and covered with grass, but there are some patches of cultivation, such as sugar, sisal-hemp, bananas and sweet potatoes.

Watched a sugar mill at work: a simple and wasteful affair of revolving toothed drums of wood turned by a walking ox; very much like the oil-crushing mills I have seen in the Sudan. The juice is collected into a wide vat and then slowly heated over a fire until a cake of pure sugar (panela) results; it is very good to eat. While it is simmering the scum is taken off, collected into gourds and left to ferment; on this the people get fearfully drunk. A few plants of tobacco are grown—not for smoking but for use in greeting. When two men meet on the road, each takes out his tobacco-box (two little gourds jammed together), hands it to the other man who opens it, dips a finger into it several times rapidly and then touches his lips with the finger. The boxes are then solemnly handed back, and each man gives the other a small pinch of green coca which he puts into his mouth and chews. ...Everyone here carries a muchila (small bag); sometimes two or three, slung over the shoulder and containing food and odds and ends. These muchilas are of roughly netted fibre or very well-woven cotton; good designs of white, yellow and purple colours. They grow a little cotton, but most of it is imported. They spin and weave a good stout cloth, and the men wear a long upper garment like an overgrown shirt slit at the sides, and a pair of short trousers. The women go in rags, relieved by strings of beads round their necks; some of the beads are beautiful carnelians, others are just pebbles—rounded and polished; occasionally red coral.

The bachelors here live in common huts together, but after marriage the man lives alone in his hut and his wife does likewise nearby. The couple eat together outside the door of the wife's hut. Children are not numerous—there are seldom more than two to a family. An attractive amusement they have is the playing upon reed instruments. They make two of different pitches—male and female—and the duets are quite pleasing though rather discordant. Mary very much taken with these pipes, and she has produced her penny whistle which is much enjoyed by them all.

JANUARY 11. Distinct earthquake about 10 p.m. last night.

Up the valley with bulls to-day to a camping place at 9000 feet. Valley steep, dry and dull; path rough, here and there a hut, and sometimes one of the so-called 'churches' inhabited by unattractive priests. They are dirty squalid places and we were not allowed to see any of the treasures said to be found in them. Almost no birds, very few butterflies, and the only conspicuous flower a bright blue lupin, beginning at about 7000 feet and growing up to 15,000.

JANUARY 12. Left our animals behind and walked up to 11,500 feet where we camped on a small platform beside a stream—the same we have followed from San Miguel. Here we found two little huts used by natives who come up to look after cattle pasturing on the 'paramos'. We pitched our Whymper tent and were snug enough. Temperature down to 33° F. in early morning and sun quite hot when out of wind.

Sun reached us at 8.30 next morning and we continued up the valley, passing at 12,700 feet a small deep lake of about

300 × 150 yards; a smaller lake drained into it from about a thousand feet above. The ground was bare granite slabs with very little vegetation. We reached the watershed of the San Miguel and Palomino valleys at 15,600 feet, and got a good view (225° from us) of a small snow peak about 17,500 feet, with small glacier on its north face. It was about half a mile from us, and as clouds were rapidly covering it and the top was too steep to attempt without ice-axe or rope, we didn't attempt to continue our climb. Rock very solid granite. Returned to San Miguel with the valley now full of clouds.

All the people of the village got very drunk last night on their sugar wine; not an attractive people. It is difficult to understand much about them, as our so-called 'interpreter' has really no knowledge of their language; but as far as I can make out, fear of sickness, fear of demons, and superstitions of one kind or another play a great part in their lives. One rather gruesome object we have seen was a large granite boulder in a particularly holy spot, and lying on it was a huge wooden club splashed with dried blood; nearby lay a flat stone with a sloping stone fixed at one end to form a chair back. This we were told was the 'divining chair' of one of the holy men. When he sits upon it he can tell the fate, or answer whatever question is put to him. This spot, particularly the murderous-looking stone, was very quickly passed by the few Indians who happened to be with us.

Back to Pueblo Viejo. Our caravan consisted of six bulls, two Indians, one half Indian, an Indian woman carrying a baby on her back, an Indian child driving a pig, a dog and ourselves.

Rode up to an Indian village called San Francisco—about 3 miles south-east—to find not a soul in the place; all away cultivating. Very disappointing, as we are hungry and had hopes of buying chickens. The Colombians hereabouts are such thieves; they ask 1½ dollars for the most miserable bundle of bones and feathers.

JANUARY 20. Left Pueblo Viejo and went up over ridge to west and down to Santa Rosa (2500 feet); then crossed the valley and camped in a side valley above a river at 3500 feet. Most lifeless region—quite destitute of living things; more like a lunar than an earthly landscape.

Next day we went steep uphill, then down and up again on to a knife-edged ridge where we found a good camping place at 6300 feet. Forest poor and dull; no flowers. Continued along the ridge to 8300 feet and got a fine view of snow peaks. Here we found a small purple rhododendron, very sticky. Down by steep zigzags until we came to Palomino valley at 2500 feet. This valley is sunnier and wider and the slopes less steep than in the narrow V-shaped gutters of the east. The village of Palomino lies on the left bank of a fine river, with beautiful pools and a good volume of water even at this dry season.

The natives refuse to accompany us up the valley and across the mountain. I had wanted to get over to the Rio Frio on the south-west of the range but must now go back to the coast....We started in the direction of the sea, north, and had a very steep climb over a ridge on a badly made path; then to a small banana clearing in the forest where we camped. We have fared badly in the matter of food on this journey and we are now both so hungry that we cannot sleep.

Down and up over innumerable spurs and foothills; heavy forest but no very big timber. Vegetation mostly dull; one lovely flower—a scarlet climber like a passion flower. Horseflies very bad. Camped by Palomino river and had a delicious bathe; water quite warm but sandflies and mosquitoes almost unbearable. Small wonder these natives don't want to come with us, for the climbing and descending of these steep ridges and the sweltering in dense jungle infested by vicious insects is no pleasant pastime.

It was a good eight hours up and down to 'Palomino by

the Sea'; a miserable spot—just a collection of palm-leaf huts set in a filthy sandy beach and infested with pigs. We borrowed the cleanest hut we could, and a swarm of visitors came through all day and all night to look at us. Our Indians won't come any further with their oxen so we must somehow fix up fresh transport here. I wish I had not relied so much on getting food locally. It is not to be had. Yet had we brought provisions with us we should have been badly held up for lack of transport, and we should not have done the half of what we have done.

JANUARY 30. Spent two days in grilling heat, and at last persuaded some Colombians to take our baggage in canoes to a place not distant along the coast. I believe these coast people, with their negro appearance, are probably descendants of escaped slaves; anyhow they are very unpleasant and none too friendly.

A long 18 miles' march mostly on the beach. Great difficulty in getting the mules along, and eventually camped at the mouth of a small river. After we had pitched our tent we went and bathed by lamplight. Got rather scared when M. swam out to the opposite bank as I had remembered seeing crocodiles not far up.

Off again along beach and then inland, south. Fine forest with wild oranges, mostly bitter but some sweet.

After eight hours' steady going we got to Calebaza, and next day we travelled down to Bonda, going over a low pass about 2000 feet into a totally different climate and vegetation. The forest damp, much more inhabited by birds and other creatures and quite unlike the burnt-up stuff on the other side. Here we satisfied our hungry bodies with beer and a fairly good luncheon; then on to Santa Marta.

We stayed in Colombia another month, and this was spent in trying to get up the Magdalena river from Barranquilla.

We were possibly foolish to trust ourselves to a little up-river steamer, but for a hundred miles or so it was plain sailing. River very low, but we could steam all day and for an hour or two after sunset, then tie up to a bank or to a 'wooding' place. Little to be seen but alligators of great size and these in enormous numbers. Several kinds of heron, terns, scissor-bills, and once or twice a pair of screamers. Lots of macaws but never a monkey.

At the end of a week our position became serious as there were only 2 feet of water ahead and our draught required three. We were tied up to a sandbank and soon joined by three other small stern-wheeler steamers. For a week we stayed on this sandbank, the river getting lower all the time. Did not see how we could get up to Bogota as the upper part of the river was now closed to navigation. Food running out and both intensely hungry; so when we sighted a small rowing boat belonging to a woodman on the bank we decided to leave the ship and start rowing down river. Stopped at various small villages on our way down, and by hiring a boat from place to place and getting a lift in a steamer where the river was deeper, we returned thankfully to Barranquilla.*

* A.F.R. read a paper before the Royal Geographical Society on 'The Sierra Nevada of Santa Marta, Colombia' which is published in the *Geographical Journal* of August 1925.

A. F. R. AT BENCOMBE, 1929

X

BACK AT KING'S

Upon our return to England we immediately set about finding a home somewhere in the West Country. A.F.R. declared that 'it must be in a spot where the rivers flow into the Atlantic; no other sea will do'. So we eventually settled at Uley, Gloucestershire, and there our children were born, and there perhaps we should be now had it not been that in the autumn of 1928 it was suggested to A.F.R. that he might return to King's as Tutor to the College.

DECEMBER 1928. (*Bencombe, Uley. To Henry Newbolt.*) You will not have heard that we are leaving this lovely place, but so we are. I am still too young to be doing nothing at 53, so an invitation to go back to Cambridge as Tutor at King's was too good to be refused, and there I go in a fortnight's time leaving my lovely family behind me....It is a big adventure again, but what has my life been but one adventure after another? and I have been simply dogged by good fortune.

(*Bencombe, Uley. To Arthur Hill.*) ...It is a serious effort to embark on a new voyage of this kind at my mature age and I have quite a number of qualms, but I have an equal number of friends on whom I shall lean heavily for support at the outset. The uprooting from here will be hateful, and the temporary separation from my family before we are installed elsewhere will be damnable—but these are small things in the long run....This comes straight from Heaven and I hope I may not be found wanting.

MARCH 1929. (*King's College. To his Wife.*) ...How lovely to see the geese flying over Bencombe. It makes one doubly glad to be alive to see such things. But you seem to be such a long way away and I keep on thinking of what you are all doing. I say to myself—Now Georgina is coming down to lunch—Now Joanna is going up to tea—and so on all the time. I simply long to come home—and damn all these rotten College meetings! However, Time does move, and the frost will soon be getting out of the ground. It seems as if the earth were softening a bit already and crocuses will soon be in flower. Young candidates for exams in and out of my rooms all the morning. Such nice lads. I really do like these young ones. They are a queer mixture of assurance and shyness—but all human and very friendly. It is a heavenly day—sun shining, and my window by my writing table wide open for the first time this year. But why am I not in the garden at home with you?

APRIL 1929. (*King's College. To his Wife.*) The sun has shone all day and the daffodils are just perfect, yet you know, dear Heart, how I wish I were home with you. To-day I must waste a lovely afternoon sitting in a room discussing fellowships—but I am not really grumbling. I like my work up here. I like the young men and I really believe they like me. I am far happier in my life here than ever I thought I should be, but it is a bit difficult for one like myself to comfortably fit in to the rather difficult community of College life.

Yesterday I borrowed a car and went miles out away into perfect country—clear chalk streams running through fields, and tree buds bursting into leaf everywhere. At least I can hear the noise of rooks from my window. Their nests are full of young ones now, so that one has to walk quickly under the trees. That reminds me, I have picked some lovely bits of Solomon's Seal and golden Kerria from the Fellows' garden and they look very pretty in the pot you gave me....

MAY 1929. (*King's College. To his Wife.*) This is the time for trippers—and they swarm everywhere, staring into my windows—foreign voices at every corner. Perhaps it is my window boxes they like so much—did I tell you of them? They are so gay. People tell me that they can see my geraniums from across the river; they are a real scarlet.

Nine aeroplanes have just gone over—flying high in a V shape like geese, but silver colour.

My darling Heart, I wish you had been with me in Chapel this afternoon. I go more often in the evening, but to-day I went in the afternoon and found it very cheering. Ord's singers sang some wonderful old music—Dowland and Byrd and Bach. The last, 'Come Jesu Come', very beautiful. I think it will be lovely for the children later on to go there instead of to the ordinary chapel service. I should like them to go to it as to a rather rare treat.

You can't think how much I long for Saturday, with you and home and our three little pigeons—and I shall smell the good smell of West Country earth, and perhaps I shall have green gooseberries!

Towards the end of May 1930 the children and I were able to leave Gloucestershire and join him in Cambridge. He was full of vigour and happiness and we delighted in being once more together. A fortnight later, on June 3, he died in his rooms at King's.

Of this tragedy, with its waste and misery, I can neither write nor speak.

Printed in the United States
By Bookmasters